T0219746

Lecture Notes in Mathematics 2201

Editors-in-Chief:
Jean-Michel Morel, Cachan
Bernard Teissier, Paris

Advisory Board:
Michel Brion, Grenoble
Camillo De Lellis, Zurich
Alessio Figalli, Zurich
Davar Khoshnevisan, Salt Lake City
Ioannis Kontoyiannis, Athens
Gábor Lugosi, Barcelona
Mark Podolskij, Aarhus
Sylvia Serfaty, New York
Anna Wienhard, Heidelberg

More information about this series at http://www.springer.com/series/304

Lars Schäfer

Nearly Pseudo-Kähler Manifolds and Related Special Holonomies

 Springer

Lars Schäfer
Institut Differentialgeometrie
Leibniz Universität Hannover
Hannover, Germany

ISSN 0075-8434   ISSN 1617-9692   (electronic)
Lecture Notes in Mathematics
ISBN 978-3-319-65806-3   ISBN 978-3-319-65807-0   (eBook)
DOI 10.1007/978-3-319-65807-0

Library of Congress Control Number: 2017951220

Mathematics Subject Classification (2010): 53C10, 53C15, 53C29, 53C30, 53C25, 53C28

Printed on acid-free paper

This Springer imprint is published by Springer Nature
The registered company is Springer International Publishing AG
The registered company address is: Gewerbestrasse 11, 6330 Cham, Switzerland

Contents

1 Introduction 1
  1.1 Stable Forms, Half-Flat Structures and Holonomy 1
  1.2 Nearly Kähler Geometry 3
  1.3 Special Kähler Manifolds 10
  1.4 Summary 11
  1.5 Zusammenfassung 13

2 Preliminaries 17
  2.1 Stable Forms 17
    2.1.1 Real Forms of $SL(3, \mathbb{C})$ 23
    2.1.2 Relation Between Real Forms of $SL(3, \mathbb{C})$ and $G_2^{\mathbb{C}}$ 25
    2.1.3 Relation Between Real Forms of $G_2^{\mathbb{C}}$ and $Spin(7, \mathbb{C})$ 27
  2.2 Almost Pseudo-Hermitian and Almost Para-Hermitian
      Geometry 28
  2.3 Linear Algebra of Three-Forms in Dimension 8 and 10 32
  2.4 Structure Reduction of Almost ε-Hermitian Six-Manifolds 34
  2.5 Pseudo-Riemannian Submersions 36
  2.6 Para-Sasaki Manifolds 37
    2.6.1 The T-Dual Space 38

3 Nearly Pseudo-Kähler and Nearly Para-Kähler Manifolds 41
  3.1 Nearly Pseudo-Kähler and Nearly Para-Kähler Manifolds 41
    3.1.1 General Properties 41
    3.1.2 Characterisations by Exterior Differential Systems
          in Dimension 6 45
    3.1.3 Curvature Identities for Nearly ε-Kähler Manifolds 50
  3.2 Structure Results 52
    3.2.1 Kähler Factors and the Structure in Dimension 8 52
    3.2.2 Einstein Condition Versus Reducible Holonomy 53
  3.3 Twistor Spaces over Quaternionic and Para-Quaternionic
      Kähler Manifolds 58

3.4   Complex Reducible Nearly Pseudo-Kähler Manifolds   62
        3.4.1   General Properties   62
        3.4.2   Co-dimension Two   63
        3.4.3   Six-Dimensional Nearly Pseudo-Kähler Manifolds   64
        3.4.4   General Dimension   66
        3.4.5   The Twistor Structure   71
3.5   A Class of Flat Pseudo-Riemannian Lie Groups   72
3.6   Classification Results for Flat Nearly ε-Kähler Manifolds   76
        3.6.1   Classification Results for Flat Nearly Pseudo-Kähler
                 Manifolds   76
        3.6.2   Classification of Flat Nearly Para-Kähler Manifolds   79
3.7   Conical Ricci-Flat Nearly Para-Kähler Manifolds   84
3.8   Evolution of Hypo Structures to Nearly Pseudo-Kähler
        Six-Manifolds   89
        3.8.1   Linear Algebra of Five-Dimensional Reductions
                 of SU(1, 2)-Structures   89
        3.8.2   Evolution of Hypo Structures   91
        3.8.3   Evolution of Nearly Hypo Structures   95
3.9   Results in the Homogeneous Case   100
        3.9.1   Consequences for Automorphism Groups   100
        3.9.2   Left-Invariant Nearly ε-Kähler Structures
                 on SL(2, $\mathbb{R}$) $\times$ SL(2, $\mathbb{R}$)   102
        3.9.3   Real Reducible Holonomy   106
        3.9.4   3-Symmetric Spaces   106
3.10   Lagrangian Submanifolds in Nearly Pseudo-Kähler Manifolds   112
        3.10.1   Definitions and Geometric Identities   113
        3.10.2   Lagrangian Submanifolds in Nearly Kähler
                 Six-Manifolds   116
        3.10.3   The Splitting Theorem   118
        3.10.4   Lagrangian Submanifolds in Twistor Spaces   123
        3.10.5   Deformations of Lagrangian Submanifolds
                 in Nearly Kähler Manifolds   127

4   Hitchin's Flow Equations   131
4.1   Half-Flat Structures and Parallel $G_2^{(*)}$-Structures   131
        4.1.1   Remark on Completeness: Geodesically Complete
                 Conformal G_2-Metrics   136
        4.1.2   Nearly Half-Flat Structures and Nearly Parallel
                 $G_2^{(*)}$-Structures   138
        4.1.3   Cocalibrated $G_2^{(*)}$-Structures and Parallel Spin(7)-
                 and $\text{Spin}_0(3, 4)$-Structures   140
4.2   Evolution of Nearly ε-Kähler Manifolds   142
        4.2.1   Cones over Nearly ε-Kähler Manifolds   142
        4.2.2   Sine Cones over Nearly ε-Kähler Manifolds   144
        4.2.3   Cones over Nearly Parallel $G_2^{(*)}$-Structures................   145

4.3   The Evolution Equations on Nilmanifolds $\Gamma \setminus H_3 \times H_3$ 146
    4.3.1   Evolution of Invariant Half-Flat Structures
        on Nilmanifolds 147
    4.3.2   Left-Invariant Half-Flat Structures on $H_3 \times H_3$ 150
    4.3.3   Solving the Evolution Equations on $H_3 \times H_3$ 157
4.4   Special Geometry of Real Forms of the Symplectic
    $SL(6, \mathbb{C})$-Module $\wedge^3 \mathbb{C}^6$ 161
    4.4.1   The Symplectic $SL(6, \mathbb{C})$-Module $V = \wedge^3 \mathbb{C}^6$
        and Its Lagrangian Cone $C(X)$
        of Highest Weight Vectors 163
    4.4.2   Real Forms (G, V_0) of the Complex Module
        $(SL(6, \mathbb{C}), V)$ 163
    4.4.3   Classification of Open G-Orbits
        on the Grassmannian X and Corresponding
        Special Kähler Manifolds 164
    4.4.4   The Homogeneous Projective Special Para-Kähler
        Manifold $SL(6, \mathbb{R}) / S(GL(3, \mathbb{R}) \times GL(3, \mathbb{R}))$ 170

References 175

Index 181

Chapter 1
Introduction

1.1 Stable Forms, Half-Flat Structures and Holonomy

Following Hitchin [76], a k-form φ on a differentiable manifold M is called *stable* if the orbit of $\varphi(p)$ under $\mathrm{GL}(T_pM)$ is open in $\Lambda^k T_p^* M$ for all $p \in M$. In this text we are mainly concerned with six-dimensional manifolds M endowed with a stable two-form ω and a stable three-form ρ. A stable three-form defines an endomorphism field J_ρ on M such that $J_\rho^2 = \varepsilon\mathrm{id}$, see (2.6). We will assume the following algebraic compatibility equations between ω and ρ:

$$\omega \wedge \rho = 0, \quad J_\rho^* \rho \wedge \rho = \frac{2}{3}\omega^3.$$

The pair (ω, ρ) defines an $\mathrm{SU}(p, q)$-structure if $\varepsilon = -1$ and an $\mathrm{SL}(3, \mathbb{R})$-structure if $\varepsilon = +1$. In the former case, the pseudo-Riemannian metric $\omega(J_\rho\cdot, \cdot)$ has signature $(2p, 2q)$. In the latter case it has signature $(3, 3)$. The structure is called *half-flat* if the pair (ω, ρ) satisfies the following exterior differential system:

$$d\omega^2 = 0, \quad d\rho = 0.$$

In [76], Hitchin introduced the following evolution equations for a time-dependent pair of stable forms $(\omega(t), \rho(t))$ evolving from a half-flat $\mathrm{SU}(3)$-structure $(\omega(0), \rho(0))$:

$$\frac{\partial}{\partial t}\rho = d\omega, \quad \frac{\partial}{\partial t}\hat{\omega} = d\hat{\rho},$$

where $\hat{\omega} = \frac{\omega^2}{2}$ and $\hat{\rho} = J_\rho^* \rho$. For compact manifolds M, he showed that these equations are the flow equations of a certain Hamiltonian system and that any solution defined on some interval $0 \in I \subset \mathbb{R}$ defines a Riemannian metric on

© Springer International Publishing AG 2017

L. Schäfer, *Nearly Pseudo-Kähler Manifolds and Related Special Holonomies*,
Lecture Notes in Mathematics 2201, DOI 10.1007/978-3-319-65807-0_1

$M \times I$ with holonomy group in G_2. We give a new proof of this theorem, which does not use the Hamiltonian system and does not assume that M is compact. Moreover, our proof yields a similar result for all three types of half-flat G-structures: $G = \mathrm{SU}(3), \mathrm{SU}(1,2)$ and $\mathrm{SL}(3, \mathbb{R})$. For the noncompact groups G we obtain a pseudo-Riemannian metric of signature $(3, 4)$ and holonomy group in G_2^* on $M \times I$ (see Theorem 4.1.3 of Chap. 4). As an application, we prove that any six-manifold endowed with a real analytic half-flat G-structure can be extended to a Ricci-flat seven-manifold with holonomy group in G_2 or G_2^*, depending on whether G is compact or noncompact, see Corollary 4.1.6 of Chap. 4. More generally, a G-structure (ω, ρ) is called *nearly half-flat* if

$$d\rho = \hat{\omega}$$

and a G_2- or G_2^*-structure defined by a three-form φ is called *nearly parallel* if

$$d\varphi = *_\varphi \varphi.$$

We prove in Theorem 4.1.12 of Chap. 4 that any solution $I \ni t \mapsto \left(\omega(t) = 2\widehat{d\rho}(t), \rho(t)\right)$ of the evolution equation

$$\frac{\partial}{\partial t}\rho = d\omega - \varepsilon\hat{\rho}$$

evolving from a nearly half-flat G-structure $(\omega(0), \rho(0))$ on M defines a nearly parallel G_2- or G_2^*-structure on $M \times I$, depending on whether G is compact or noncompact, (see (2.3) of Chap. 4) for the definition of $\widehat{d\rho}$. For compact manifolds M and $G = \mathrm{SU}(3)$ this theorem was proven by Stock [114].

The above constructions are illustrated in Sect. 4.2 of Chap. 4, where we start with a nearly pseudo-Kähler or a nearly para-Kähler six-manifold as initial structure. These structures are both half-flat and nearly half-flat and the resulting parallel or nearly parallel G_2- and G_2^*-structures induce cone or (hyperbolic) sine cone metrics.

In Sect. 4.3 of Chap. 4, we discuss the evolution of invariant half-flat structures on nilmanifolds. Lemma 4.3.1 of Chap. 4 shows how to simplify effectively the ansatz for a solution for a number of nilpotent Lie algebras including the direct sum $\mathfrak{g} = \mathfrak{h}_3 \oplus \mathfrak{h}_3$ of two Heisenberg algebras. Focusing on this case, we determine the orbits of the $\mathrm{Aut}(\mathfrak{h}_3 \oplus \mathfrak{h}_3)$-action on non-degenerate two-forms ω on $\mathfrak{g}$ which satisfy $d\omega^2 = 0$. Based on this, we describe all left-invariant half-flat structures (ω, ρ) on $H_3 \times H_3$. A surprising phenomenon occurs in indefinite signature. Under the assumption $\omega(\mathfrak{z}, \mathfrak{z}) = 0$, which corresponds to the vanishing of the projection of ω on a one-dimensional space, the geometry of the metric induced by a half-flat structure (ω, ρ) is completely determined (Proposition 4.3.7 of Chap. 4) and the evolution turns out to be affine linear (Proposition 4.3.10 of Chap. 4). However, this evolution produces only metrics that are decomposable and have one-dimensional holonomy group. On the other hand, we give an explicit formula in Proposition 4.3.12 of Chap. 4 for the parallel three-form φ resulting from the evolution for any half-flat structure (ω, ρ) with $\omega(3, 3) \neq 0$. In fact, the formula is completely algebraic such that the integration of the differential equation is circumvented. In particular, we give a number of explicit examples of half-flat structures of the second kind on $\mathfrak{h}_3 \oplus \mathfrak{h}_3$ which evolve to new metrics with holonomy group equal to G_2 and G_2^*. Moreover, we construct an eight-parameter family of half-flat deformations of the half-flat examples which lift to an eight-parameter family of deformations of the corresponding parallel stable three-forms in dimension 7. Needless to say, those examples of $G_2^{(*)}$-metrics on $M \times (a, b)$ for which $(a, b) \neq \mathbb{R}$ are geodesically incomplete. However, for M compact with an SU(3)-structure, a conformal transformation produces complete Riemannian metrics on $M \times \mathbb{R}$ that are conformally parallel G_2.

A G_2- or G_2^*-structure defined by a three-form φ is called *cocalibrated* if

$$d *_\varphi \varphi = 0.$$

Hitchin proposed the following equation for the evolution of a cocalibrated G_2-structure $\varphi(0)$:

$$\frac{\partial}{\partial t}(*_\varphi \varphi) = d\varphi.$$

He proved that any solution $I \ni t \mapsto \varphi(t)$ on a compact manifold M defines a Riemannian metric on $M \times I$ with holonomy group in Spin(7). We generalise also this theorem to noncompact manifolds and show that any solution of the evolution equation starting from a cocalibrated G_2^*-structure defines a pseudo-Riemannian metric of signature $(4, 4)$ and holonomy group in $\text{Spin}_0(3, 4)$, see Theorem 4.1.13 of Chap. 4.

1.2 Nearly Kähler Geometry

Nearly Kähler manifolds have been introduced and studied by Gray [64–67] in a series of papers in the seventies in the context of weak holonomy. These manifolds are almost Hermitian manifolds which are of type $\mathcal{W}_1$ in the well-known Gray-Hervella classification [68] of almost Hermitian structures. The most prominent example is the six-sphere endowed with the round metric and the almost complex structure induced by the cross product coming from the octonions [62, 63]. New interest in the subject came in the beginning of the nineties from the relation between real Killing spinors on six-manifolds and nearly Kähler manifolds given by Grunewald [70]. Shortly after this paper the classification of manifolds admitting real Killing spinors was given in [12] by using a cone construction. More precisely, in this reference it is shown that a real Killing spinor on M corresponds to a parallel spinor on the metric cone over M. By a theorem of Gallot [60] it follows, that, if the holonomy of the cone $\hat{M}$ (over a compact simply connected manifold M) is reducible, then $\hat{M}$ is flat and M is isometric to the standard sphere. Excluding the sphere the reference [12] obtains a list of possible holonomies of the cone over manifolds admitting real Killing spinors and analyses the induced geometry on M. Let us observe, that in the semi-Riemannian setting the situation is different, since Section 3 of [8] gives a number of examples, where the statement of Gallot's theorem is not true. Nonetheless cone constructions are useful in the semi-Riemannian setting and are still an area of active research in Riemannian geometry. For example, inspired by the classical correspondence between Ricci-flat Kähler manifolds and Sasaki-Einstein manifolds (see Section 5 of [12] or Chapter 11.1 of [20]) the recent article [35] studies this type of correspondence in the setting of 'geometries with torsion'. More recently, the reference [1] gives a systematic approach to cone constructions over manifolds carrying a G structure endowed with a 'characteristic connection', i.e. a metric connection with skew symmetric torsion preserving the G structure and obtains 'torsionful' generalisations of the above mentioned results. In the same way nearly Kähler geometry is an area of active research during the last decades [16, 25, 98, 99]. The classification of complete simply connected strict nearly Kähler manifolds was reduced to dimension 6 by Nagy [98, 99] using a result of Cleyton and Swann [34]. Butruille [25] has shown under the same assumptions that all homogenous strict nearly Kähler manifolds are 3-symmetric. The first examples of complete inhomogeneous strict nearly Kähler manifolds had to wait until the beginning of this year [55]! For interest and applications of all these geometric structures in physics, namely supergravity and string theories, we may refer to [49, 57, 115].

To our best knowledge only Gray considers pseudo-Riemannian metrics in his paper on 3-symmetric spaces [66]. The notion of a *nearly para-Kähler manifold* was recently introduced by Ivanov and Zamkovoy [78] and originates in the study of Killing spinors in semi-Riemannian geometry by Kath [81, 82]. These manifolds naturally appear as one class in the classification of almost para-Hermitian manifolds [59, 86], which generalises the Gray-Hervella list. In contrast to Riemannian geometry, where a Ricci-flat nearly Kähler manifold is automatically a Kähler manifold (compare Remark 3.1.9), a Ricci-flat nearly para-Kähler manifold needs not necessarily to be para-Kähler (In fact, it is not, as a posteriori shown by our examples). Therefore Ivanov and Zamkovoy [78] ask for examples of this type and the same authors observe, that at this time there were no known examples of Ricci-flat (non-Kähler) nearly para-Kähler manifolds in the literature.

In addition, nearly para-Kähler metrics (or more generally almost para-Hermitian metrics) are of split signature, i.e. signature (n, n) and for this reason of special interest in the study of semi-Riemannian holonomy [17]. More recently nearly para-Kähler geometry is of interest in the setting of projective holonomy [61].

The interest in flat nearly pseudo-Kähler manifolds is also motivated by the study of tt^*-structures (topological-antitopological fusion structures) on the tangent-bundle. In fact, flat nearly pseudo-Kähler manifolds provide a class of tt^*-structures on the tangent-bundle [107]. A second important class of solutions is given by *special Kähler manifolds* [40]. In other words, one can interpret tt^*-structures on the tangent-bundle as a common generalisation of these two geometries. Therefore we shortly review this subject.

Topological-antitopological fusion or tt^*-geometry is a topic of mathematical and physical interest. In about 1990 physicists studied topological-field-theories and their moduli spaces, in particular $N = 2$ supersymmetric field-theories and found a special geometric structure called topological-antitopological fusion (see the works of Cecotti and Vafa [30] and Dubrovin [48]). These geometries are realised on the tangent bundle of some manifold and part of their data is a Riemannian metric. One can replace the tangent bundle by an abstract vector bundle. This step allows for instance to consider tt^*-bundles as a generalisation of variations of Hodge structures, as it was done in Hertling's paper [74].

Let us now shortly recall some notions of tt^*-geometry and refer to [42] for some survey on this differential geometric viewpoint of the field. A tt^*-bundle (E, D, S) over an almost (para-)complex manifold (M, J) is a real vector bundle $E \rightarrow M$ endowed with a connection D and a section $S \in \Gamma(T^*M \otimes \mathrm{End}\, E)$ which satisfy the tt^*-equation

$$R^\theta = 0 \quad \text{for all} \quad \theta \in \mathbb{R}, \tag{1.1}$$

where R^θ is the curvature tensor of the connection D^θ defined by

$$D_X^\theta := D_X + \cos(\theta)S_X + \sin(\theta)S_{JX} \quad \text{for all} \quad X \in TM. \tag{1.2}$$

A metric tt^*-bundle (E, D, S, g) is a tt^*-bundle (E, D, S) endowed with a possibly indefinite D-parallel fibre metric g such that for all $p \in M$

$$g(S_X Y, Z) = g(Y, S_X Z) \quad \text{for all} \quad X, Y, Z \in T_p M. \tag{1.3}$$

A symplectic tt^*-bundle (E, D, S, ω) is a tt^*-bundle (E, D, S) endowed with the structure of a symplectic vector bundle (E, ω), such that ω is D-parallel and S is ω-symmetric, i.e. for all $p \in M$

$$\omega(S_X Y, Z) = \omega(Y, S_X Z) \quad \text{for all} \quad X, Y, Z \in T_p M. \tag{1.4}$$

In [106, 107] we have shown the following general construction principle of such geometries.

Theorem 1.2.1 *Given an almost (para-)complex manifold (M, J) with a flat connection ∇ and a decomposition of $\nabla = D + S$ in a connection D and a section S in $T^*M \otimes \mathrm{End}\,(TM)$, such that J is D-parallel, i.e. $DJ = 0$. If (TM, D, S) defines a tt^*-bundle, such that D^θ is $\nabla^{\alpha\theta}$ with $\alpha = \pm 2$, then D and S are uniquely determined by $S = -\frac{1}{2}J^{-1}(\nabla J)$ and $D = \nabla - S$.*

Moreover, (TM, D, S) as above defines a tt^*-bundle, such that D^θ is $\nabla^{\alpha\theta}$ with $\alpha = \pm 2$, if and only if J satisfies

$$(\nabla_{JX} J) = \pm J(\nabla_X J) \tag{1.5}$$

and D and S are given by $S = -\frac{1}{2} J^{-1}(\nabla J)$ and $D = \nabla - S$.

As an application this theorem gives solutions of symplectic tt^*-bundles on the tangent bundle, which are more general than nearly Kähler manifolds in the sense, that we admit connections ∇ having torsion, but more special in the sense, that our connection ∇ has to be flat:

Theorem 1.2.2 *Given an almost (para-)Hermitian manifold (M, J, g) with a flat metric connection ∇, such that (∇, J) satisfies the nearly (para-)Kähler condition. Define S, a section in $T^*M \otimes \mathrm{End}\,(TM)$ by*

$$S := -\frac{1}{2} J^{-1}(\nabla J), \tag{1.6}$$

then $(TM, D = \nabla - S, S, \omega = g(\cdot, J^{-1}\cdot))$ defines a symplectic tt^*-bundle. Moreover, it holds $DJ = 0$ and $T^D = T^\nabla - 2S$.

At the moment, when the present author has found this new class of solutions to the tt^*-equations, there were no known examples of Levi-Civita flat nearly para-Kähler manifolds. This is the starting point of our interest in the subject. As one of the results presented in the present text we discuss the classification of these manifolds in Sect. 3.6 of Chap. 3. As mentioned above, at about the same time Ivanov and Zamkovoy [78] asked for examples of Ricci-flat nearly para-Kähler manifolds in six dimensions with $DJ \neq 0$. As a consequence of our results we show that there exists compact six-dimensional such manifolds with $DJ \neq 0$ (see Sect. 3.6 of Chap. 3).

Another class of almost Hermitian manifolds solving the condition (1.5) is special Kähler manifolds, which shows, that there is a strong similarity to special Kähler geometry. This duality can also be seen in the present work. We construct flat nearly pseudo-Kähler manifolds of split signature from a certain constant three-form, while in [14] flat special Kähler manifolds were constructed from a symmetric three-tensor. The classification results are given in Sect. 3.6 of Chap. 3. The first one is Theorem 3.6.4, which encodes a flat nearly pseudo-Kähler structure in a constant three-form η subject to two constraints. An explicit formula for J in terms of η is given. Next we analyse the constraints on η. It turns out that the first is equivalent to require that η has isotropic support (cf. Proposition 3.6.5) and the second is equivalent to a type condition on η. We explicitly solve the two constraints on the real three-form η (in $4m$ variables) in terms of a freely specifiable complex three-form $\zeta \in \Lambda^3(\mathbb{C}^m)^*$. In particular, any such form $\zeta \neq 0$ defines a complete simply connected strict flat nearly pseudo-Kähler manifold, see Corollary 3.6.11.

Further we show that any strict flat nearly pseudo-Kähler manifold is locally the product of a flat pseudo-Kähler factor of maximal dimension and a strict flat nearly pseudo-Kähler manifold of dimension $4m \geq 12$ and split signature (Theorem 3.6.9). This implies, in particular, the non-existence of strict flat nearly Kähler manifolds with positive definite metric. This subsection finishes with the classification of complete simply connected flat nearly Kähler manifolds up to isomorphism in terms of $GL_m(\mathbb{C})$-orbits on $\Lambda^3(\mathbb{C}^m)^*$, see Corollary 3.6.12.

This motivates our excursion to flat structures on Lie groups: Let V be a pseudo-Euclidian vector space and $\eta \in \Lambda^3 V$. Contraction with η defines a linear map $\Lambda^2 V^* \to V$. The image is the support of η. In Sect. 3.5 we show that any 3-vector $\eta \in \Lambda^3 V$ with isotropic support defines a simply connected 2-step nilpotent Lie group $\mathcal{L}(\eta)$ with a flat bi-invariant pseudo-Riemannian metric h of the same signature as V. We prove that this exhausts all simply connected Lie groups with a flat bi-invariant metric, see Theorem 3.5.2. Let $\dim V$ be even and fix $J \in \mathfrak{so}(V)$, such that $J^2 = -Id$ or $J^2 = Id$. We denote the corresponding left-invariant endomorphism field on the group $\mathcal{L}(\eta)$ again by J. Assume that $\eta \in \Lambda^3 V$ has isotropic support and has, in addition, type $(3, 0) + (0, 3)$. Then $(\mathcal{L}(\eta), h, J)$ is a flat nearly Kähler manifold if $J^2 = -Id$ and a flat nearly para-Kähler manifold if $J^2 = Id$. This follows from our classification results (Theorems 3.6.4 and 3.6.17). Moreover it is shown that any complete simply connected flat nearly (para-)Kähler manifold is of this form. To sum up, we have shown that any simply connected complete flat nearly (para-)Kähler manifold is a Lie group $\mathcal{L}(\eta)$ with a left-invariant nearly (para-)Kähler structure and bi-invariant metric. Conversely, it follows from unpublished work of Cortés and Nagy, that a Lie group with a left-invariant nearly (para-)Kähler structure and bi-invariant metric is necessarily flat and is therefore covered by one of our groups $\mathcal{L}(\eta)$. The proof of this statement uses the unicity of the connection with totally skew-symmetric torsion preserving the nearly (para-)Kähler structure and the Jacobi identity. Further, suppose that $\Gamma \subset \mathcal{L}(\eta)$ is a lattice. Then the almost (para-)complex structure J on the group $\mathcal{L}(\eta)$ induces an almost (para-)complex structure J on the compact manifold $M = M(\eta, \Gamma) = \Gamma \backslash \mathcal{L}(\eta)$. Therefore (M, h, J) is a compact nearly (para-)Kähler manifold. However, the (para-)complex structure is not $\mathcal{L}(\eta)$-invariant, unless $\eta = 0$. Moreover, for $\eta \neq 0$, (M, h, J) is an inhomogeneous nearly (para-)Kähler manifold, that is, it does not admit any transitive group of automorphisms of the nearly (para-)Kähler structure. Since J is not right-invariant, this follows from the fact that $\mathrm{Isom}_0(M)$ is obtained from the action of $\mathcal{L}(\eta)$ by right-multiplication on M, see Corollary 3.5.6. The first such non-trivial flat compact nearly para-Kähler nilmanifold $M(\eta) = \Gamma \backslash \mathcal{L}(\eta)$ is six-dimensional and is obtained from a non-zero element $\eta \in \Lambda^3 V^+ \cong \mathbb{R}$, where $V^+ \subset V = \mathbb{R}^{3,3}$ is the $+1$-eigenspace of J.

The next natural question is the existence of non-flat Ricci-flat examples of strict (i.e. non-Kähler) nearly para-Kähler six-manifolds. Let us shortly describe the construction we give in Sect. 3.7: Starting with a Ricci-flat nearly para-Kähler manifold (M, τ, g), we make an ansatz P depending on τ and a vector field ξ. This P is a second almost para-complex structure if the Nijenhuis tensor of τ has isotropic support (see Definition 2.3.1) and is para-Kähler if ξ is a conical vector field in the sense of Definition 3.7.1. Namely, we prove in Theorem 3.7.3 that the data (M, P, g, ξ) is a conical Ricci-flat para-Kähler manifold and give the inverse of this construction in Theorem 3.7.5. Space-like regular conical Ricci-flat para-Kähler manifolds induce the structure of a para-Sasaki Einstein manifold on the level set $\{g(\xi, \xi) = c^2 \neq 0\}$, cf. Proposition 3.7.8. As an application of our correspondence we obtain the following result:

Theorem 1.2.3 Let (N^5, g, T) be a para-Sasaki Einstein manifold of dimension 5 and denote by $(M^6 = \widehat{N}, \widehat{g}, P, \xi)$ the associated conical Ricci-flat para-Kähler manifold on the cone $M = \widehat{N}$ over N. Then the cone M can be endowed with the structure of a conical Ricci-flat strict nearly para-Kähler six-manifold $(M, \tau, \widehat{g}, \xi)$. Moreover, M is flat if and only if N has constant curvature.

Further we like to mention that starting this construction with five-dimensional para-Sasaki Einstein manifolds of non-constant curvature yields Ricci-flat nearly para-Kähler six-manifolds, which are not flat and hence were not known in the literature before. Different classes of five-dimensional para-Sasaki Einstein manifolds are given in Sect. 2.6 of Chap. 2. The examples obtained from the present method cannot be compact. Compact examples are only known by the above described results for the flat case. To our knowledge the existence of compact (non-flat) Ricci-flat examples remains still open.

A well-known characterisation [103] of a nearly Kähler structure on a six-manifold is the one as an SU(3)-structure (ω, ψ^+, ψ^-) satisfying the exterior system

$$d\omega = 3\psi^+, \tag{1.7}$$

$$d\psi^- = \nu\, \omega \wedge \omega \tag{1.8}$$

for a real constant ν which depends on sign and normalisation conventions.

As a consequence, nearly Kähler manifolds are for example a special class of half-flat structures.

There exists a left-invariant nearly Kähler structure on $S^3 \times S^3$ which arises from a classical construction of 3-symmetric spaces by Ledger and Obata [90] and this exterior system is the main tool in the proof, that this nearly Kähler structure is *the only one* on $S^3 \times S^3$ up to homothety ([25], compare also the extended English version [26]). In fact, the proof of this uniqueness result has been the most difficult step in the classification of homogeneous nearly Kähler structures in dimension 6 [25, 26].

This Riemannian remark is the starting point of the following development. The construction of a 3-symmetric space from $G = \mathrm{SL}(2, \mathbb{R})$ instead of SU(2) defines a left-invariant nearly pseudo-Kähler structure on $\mathrm{SL}(2, \mathbb{R}) \times \mathrm{SL}(2, \mathbb{R})$. We shortly recall this construction explicitly. The group $G \times G \times G$ admits a symmetry of order three given by $(g_1, g_2, g_3) \mapsto (g_2, g_3, g_1)$ which stabilises the diagonal Δ. The tangent space of $M^6 = G \times G \times G / \Delta$ is identified with

$$\mathfrak{p} = \{(X, Y, Z) \in \mathfrak{g} \oplus \mathfrak{g} \oplus \mathfrak{g} \mid X + Y + Z = 0\}.$$

Denote by $K_\mathfrak{g}$ the Killing form of $\mathfrak{g}$ and define an invariant scalar product on $\mathfrak{g} \oplus \mathfrak{g} \oplus \mathfrak{g}$ by $g = K_\mathfrak{g} \oplus K_\mathfrak{g} \oplus K_\mathfrak{g}$. This yields a naturally reductive metric on M^6. Using Proposition 5.6 of [66] this metric is nearly pseudo-Kähler. For completeness sake we recall that the complex structure is given by

$$J(X, Y, Z) = \frac{2}{\sqrt{3}}(Z, X, Y) + \frac{1}{\sqrt{3}}(X, Y, Z).$$

Considering Butruille's results, it is natural to ask how many left-invariant nearly pseudo-Kähler structures there are on $\mathrm{SL}(2, \mathbb{R}) \times \mathrm{SL}(2, \mathbb{R})$. Comparing with the results mentioned in the last paragraph, the answer seems a priori hard to guess. The result of Sect. 3.9.2 of Chap. 3 is the proof that there is a unique left-invariant nearly pseudo-Kähler structure on all Lie groups with Lie algebra $\mathfrak{sl}(2, \mathbb{R}) \oplus \mathfrak{sl}(2, \mathbb{R})$. A byproduct of the proof is the result that there are no nearly para-Kähler structures on these Lie groups. We add the remark that there exist co-compact lattices for these Lie groups. Indeed, the article [102] contains a complete list of the compact quotients of Lie groups with Lie algebras $\mathfrak{sl}(2, \mathbb{R})$, which also give rise to compact quotients on a direct product of such groups.

Another important application of the exterior system is the evolution of hypo-structures to Calabi-Yau metrics by the Conti-Salamon evolution equations [36] and that of nearly hypo-structures to nearly Kähler structures [53]. We extend these constructions for all possible signatures in Sect. 3.8 of Chap. 3.

Let us now shortly sum up some of our structure results for pseudo-Riemannian signature: A general condition to split off the Kähler factor from a nearly pseudo-Kähler manifold is obtained in Theorem 3.2.1. A strict nearly pseudo-Kähler six-manifold M^6 is shown to be Einstein in Theorem 3.2.8. An application of Theorem 3.2.1 implies after using some linear algebra of three-forms the splitting result (Theorem 3.2.4) for *nice* nearly pseudo-Kähler manifolds (cf. Definition 3.2.3) in dimension 8. The argument also holds true for a Riemannian metric and gives an alternative proof for the known result in positive signature.

If a *nice* nearly pseudo-Kähler ten-manifold is in addition *decomposable* (cf. Definitions 3.2.3 and 3.2.6), we find two cases:

(1) In the first we can split off the Kähler factor and
(2) in the second the holonomy of $\bar{\nabla}$ is reducible with a complex one dimensional factor.

The situation in (2) is one motivation to study twistor spaces: An interesting class of nearly Kähler manifolds M^{4n+2} can be constructed from twistor spaces over positive quaternionic Kähler manifolds. This class is characterised [16, 98] by the reducibility of the holonomy of the canonical connection $\bar{\nabla}$ to $U(n) \times U(1)$. We show in Sect. 3.3 of Chap. 3 that the twistor spaces over negative quaternionic Kähler manifolds and para-quaternionic Kähler manifolds carry a nearly pseudo-Kähler structure and characterise the class of such examples by a holonomic condition in Sect. 3.4 of Chap. 3. In the pseudo-Riemannian setting twistor spaces are a good source of examples, since quaternionic geometry is richer in negative scalar curvature than in positive (cf. Remark 3.3.7) and since we have the additional class of twistor spaces over para-quaternionic manifolds. In the last section we prove that a nearly pseudo-Kähler manifold M of *twistorial type* (cf. Definition 3.4.11) is obtained from the above mentioned construction on a twistor space. This is done as follows: We prove that M comes from a pseudo-Riemannian submersion $\pi : M \to N$. Then we use the nearly Kähler data on M to endow N with the structure of a (para-)quaternionic manifold. The proof is finished by identifying the twistor space of N with M. The former proofs [16, 98] in the Riemannian case all use the inverse twistor construction of Penrose or LeBrun, which does not seem to be developed for the situations occurring in this text. As the reader might observes, the approach presented here holds also true for Riemannian metrics. As an application of our results we get the following information:

Theorem 1.2.4 *Let (M^{10}, J, g) be a nice decomposable nearly pseudo-Kähler manifold, then the universal cover of M is either the product of a pseudo-Kähler surface and a (strict) nearly pseudo-Kähler manifold M^6 or a twistor space of an eight-dimensional (para-)quaternionic Kähler manifold endowed with its canonical nearly pseudo-Kähler structure.*

Since the fundamental two-form of a nearly Kähler manifold is not closed, Lagrangian or maximal totally real submanifolds are more rigid. As we shall see in the Sect. 3.10.2 these submanifolds are minimal, or more precisely have vanishing mean curvature in dimension 6. Moreover, we prove in Theorem 3.10.10 of Sect. 3.10.3, that such submanifolds split according to the decomposition of the ambient space into its Kähler and strict nearly Kähler factor and apply this information to obtain more precise structure results in dimension 8 and 10. In the third subsection we show that for the class of nearly Kähler spaces coming from the twistor construction Lagrangian submanifolds are also minimal. The last subsection studies deformations of Lagrangian submanifolds.

1.3 Special Kähler Manifolds

As mentioned above special Kähler manifolds provide a class of tt^*-bundles which is closely related to nearly Kähler geometry. Homogeneous projective special pseudo-Kähler manifolds of semisimple groups with compact stabiliser were classified in [4]. It follows that there is a unique homogeneous projective special pseudo-Kähler manifold with compact stabiliser which admits a transitive action of a real form of $SL(3, \mathbb{C})$ by automorphisms of the special Kähler structure, namely

$$\frac{SU(3,3)}{S(U(3) \times U(3))}.$$

Its special Kähler metric is (negative) definite. The above manifold occurred in [4] as an open orbit of SU(3, 3) on the projectivised highest weight vector orbit of SL(6, $\mathbb{C}$) on $\Lambda^3(\mathbb{C}^6)^*$. The space of stable three-forms $\rho \in \Lambda^3(\mathbb{R}^6)^*$, such that $J_\rho^2 = -1$, has also the structure of a special pseudo-Kähler manifold [76]. The underlying projective special pseudo-Kähler manifold is the manifold

$$\frac{\text{SL}(6, \mathbb{R})}{\text{U}(1) \cdot \text{SL}(3, \mathbb{C})}$$

which has noncompact stabiliser and indefinite special Kähler metric. Both manifolds can be obtained from the space of stable three-forms $\rho \in \Lambda^3(\mathbb{C}^6)^*$ by imposing two different reality conditions. In the last section of this text we determine all homogeneous spaces which can be obtained in this way and describe their special geometric structures. In particular, we calculate the signature of the special Kähler metrics. For the projective special pseudo-Kähler manifold SL(6, $\mathbb{R}$)/ (U(1) · SL(3, $\mathbb{C}$)), for instance, we obtain the signature (6, 12). Apart from the two above examples, we find two additional special pseudo-Kähler manifolds and also a special para-Kähler manifold. The latter is associated to the space of stable three-forms $\rho \in \Lambda^3(\mathbb{R}^6)^*$, such that $J_\rho^2 = +1$.

1.4 Summary

Let us shortly present the structure of this text. Chapter 2 collects some basics and starts with an introduction to the formalism of stable forms in which we thoroughly discuss its application to the different real forms of SL(3, $\mathbb{C}$), $G_2^{\mathbb{C}}$ and Spin(7, $\mathbb{C}$). The second section focuses on the basic notions of (para-)Hermitian geometry and $U^\varepsilon(p, q)$-structures and the third prepares some linear algebra of three forms, which is needed in the third chapter to construct the above cited examples of Ricci-flat nearly (para-)Kähler manifolds and to obtain the splitting results. The next section of this chapter recalls some geometry of pseudo-Riemannian submersions. In the last section of the first chapter we recall some notions on para-Sasaki geometries and discuss the construction method given by T-duality.

In the third chapter we consider nearly para-Kähler manifolds. Starting with the definition of a nearly ε-Kähler manifold (Definition 3.1.1) we generalise its characterisation in terms of $d\omega$ and the Nijenhuis tensor (Proposition 3.1.2). Next we introduce the characteristic connection $\bar{\nabla}$ of an almost ε-Hermitian manifold (M, J, g, ω) (Theorem 3.1.5). We point out the fact, that this connection has parallel torsion (Proposition 3.1.7) and face for the first time the observation that one cannot apply the stable form technique in the Ricci-flat setting (cf. Proposition 3.1.10), since the support of the Nijenhuis tensor is isotropic. The first section ends deducing some helpful consequences for the setting of isotropic Nijenhuis tensor. The second subsection is devoted to the proof of the exterior differential system characterising those $\text{SU}^\varepsilon(p, q)$-structures (ω, ψ^+, ψ^-) which correspond to nearly

Kähler manifolds (Theorem 3.1.19). On the way to this result information on more general exterior systems is derived. The last subsection collects a number of curvature identities for nearly ε-Kähler manifolds. In the second section we treat structure results. More precisely, we split off the Kähler factor (Theorem 3.2.1), give the relation between the Einstein condition and reducible holonomy (Theorem 3.2.5) for the characteristic connection and apply these to the observations in dimension 8 and 10 (Theorems 3.2.4 and 3.2.10). One of the classes motivates to consider the twistorial construction of nearly ε-Kähler spaces, which we discuss in detail in the third section of this chapter. Since for these examples the characteristic connection has complex reducible holonomy, it is a natural question to characterise the class of nearly ε-Kähler spaces having this property in Sect. 3.4. After deriving some information in the first two subsections, we focus on the six-dimensional case, which illustrates the necessary geometric assumption, called *of twistorial type*, for the general case of the fourth subsection. The last subsection combines the information to prove the inverse twistor construction in Theorem 3.4.18. The next section is an excursion to our classification of flat pseudo-Riemannian Lie groups (Theorem 3.5.1). Combined with our classification results for flat nearly ε-Kähler manifolds of the sixth subsection this yields compact such examples. These results impose the question to find (non-flat, non-Kähler) Ricci-flat examples. The positive answer is a correspondence between conical Ricci-flat para-Kähler and conical Ricci-flat strict nearly para-Kähler manifolds in Theorems 3.7.3 and 3.7.5 of the present chapter. This construction has an interpretation as a cone construction over para-Sasaki Einstein manifolds (Theorem 3.7.9). Remaining in the area of co-dimension 1 constructions we generalise as another application of the above cited exterior system the evolution of hypo and nearly hypo structures to the pseudo-Riemannian setup in Sect. 3.8. As further application of the exterior system we observe in Sect. 3.9.1, that the automorphisms of certain classes of $SU^{\varepsilon}(p,q)$-structures (ω, ψ^{+})—including nearly Kähler structures—are completely described by demanding, that these fix only one of the two stable forms. In Sect. 3.9.2, we finally obtain the aforementioned structure results on $SL(2,\mathbb{R}) \times SL(2,\mathbb{R})$ and we also extend the results on $S^3 \times S^3$ by proving the non-existence of nearly (para-)Kähler structures of indefinite signature. The following two subsections consider real reducible holonomy of the characteristic connection and the image of 3-symmetric spaces under T-duality. The subject of the last section of the third chapter is Lagrangian submanifolds. In the first subsection we derive some geo-metric information on Lagrangian submanifolds of nearly Kähler spaces, which are used in the second section to prove, that in dimension 6 such manifolds are minimal, i.e. have vanishing mean curvature. In the subsequent section we show, that a Lagrangian submanifold splits according to the decomposition of the ambient nearly Kähler manifolds into its Kähler and strict nearly Kähler factor. The fourth subsection considers ambient nearly Kähler spaces coming from the twistor construction and the last subsection is devoted to deformations of Lagrangian submanifolds of nearly Kähler spaces.

The fourth chapter studies the above cited extension of Hitchin's flow equations. In the first subsection we give the lift from half-flat structures to metrics with holonomy $G_2^{(*)}$ (Theorem 4.1.3), discuss completeness of the corresponding metrics and extend co-calibrated $G_2^{(*)}$ to parallel Spin(7)- and $\text{Spin}_0(3, 4)$-structures. The second subsection then evolves nearly Kähler structures to (nearly) parallel $G_2^{(*)}$-structures. The sine-cones over these $G_2^{(*)}$-structures then yield metrics of holonomy Spin(7)- and $\text{Spin}_0(3, 4)$. In Sect. 1.3 we give explicit examples on nilmanifolds of the type $\Gamma\backslash(H_3 \times H_3)$. In the first subsection we reduce the evolution of $\hat{\omega}$ to some 4-dimensional subspace and classify left-invariant half-flat structures. One class is rather rigid, i.e. either flat or a product of a para-hyper-Kähler symmetric space N^4 and a flat factor (cf. Proposition 4.3.7). After these preparations we can solve the evolution equations for all the classes of invariant half-flat structures and find examples of full holonomy G_2 and G_2^* (Theorem 4.3.14). The last subsection collects the above announced results on special Kähler geometry.

1.5 Zusammenfassung

Das zweite Kapitel der Arbeit stellt einige später benötigte Grundlagen zusammen. Es beginnt mit einer Einführung in den Formalismus der stabilen Formen, wobei wir im Detail dessen Anwendung im Falle der verschiedenen reellen Formen der Lieschen Gruppen $SL(3, \mathbb{C})$, $G_2^{\mathbb{C}}$ und $\text{Spin}(7, \mathbb{C})$ ausarbeiten. Im zweiten Abschnitt besprechen wir (para-)hermitesche Geometrien und $U^\varepsilon(p, q)$-Strukturen, während der dritte Abschnitt etwas lineare Algebra von Differentialformen dritter Stufe vorbereitet, welche im folgenden Verlauf verwandt wird, um die oben erwähnten Zerlegungssätze zu erhalten und Ricci-flachen nearly para-Kähler-Beispiele zu konstruieren. Bevor der letzte Abschnitt an einige Definitionen und Grundlagen zu para-Sasaki-Mannigfaltigkeiten erinnert und die Konstruktion solcher Geometrien mittels T-Dualität erläutert, wiederholt der nächste Abschnitt kurz etwas Geometrie von pseudo-Riemannschen Submersionen.

   Das dritte Kapitel befaßt sich mit nearly (para-)Kähler-Mannigfaltigkeiten. Beginnend mit der Definition einer nearly ε-Kähler-Mannigfaltigkeit (Definition 3.1.1) verallgemeinern wir die Charakterisierung dieser Klasse von fast-ε-hermiteschen Geometrien in Termen des Nijenhuis-Tensors und von $d\omega$, wobei ω die fundamentale 2-Form der unterliegenden fast-hermiteschen Struktur ist (Proposition 3.1.2). Im Anschluß führen wir den charakteristischen Zusammenhang $\bar{\nabla}$ einer fast-ε-hermiteschen Mannigfaltigkeit ein (Theorem 3.1.5) und zeigen, dass dieser parallele Torsion hat (vgl. Proposition 3.1.10). An dieser Stelle begegnen wir zum ersten Mal der Tatsache, dass der Formalismus stabiler Formen nicht in der Ricci-flachen Situation anwendbar ist (cf. Propostion 3.1.10), weil der Support des Nijenhuis-Tensors hier isotrop ist. Wir beenden daher den ersten Abschnitt mit der Herleitung nützlicher Informationen für Geometrien mit isotropem Nijenhuis-Tensor. Im zweiten Abschnitt liefern wir nun den Beweis des äußeren Systems, welches die zu nearly Kähler-Mannigfaltigkeiten gehörigen $SU^\varepsilon(p, q)$-Strukturen (ω, ψ^+, ψ^-) charakterisiert (Theorem 3.1.19). Auf dem Weg zu diesem Resultat leiten wir Informationen zu allgemeineren äußeren Systemen her, bevor wir im letzten Unterabschnitt verschiedene Kümmungsidentitäten für nearly ε-Kähler-Mannigfaltigkeiten sammeln. Im zweiten Abschnitt behandeln wir die oben beschriebenen Strukturresultate. Zuerst splitten wir den Kähler-Faktor ab (Theorem 3.2.1), stellen einen Bezug zwischen der Einstein-Bedingung und reduzibler Holonomie des charakteristischen Zusammenhangs her (Theorem 3.2.5) und wenden dies in den Dimensionen acht und zehn an (Theorem 3.2.4 und Theorem 3.2.10). Eine der dabei erhaltenen Klassen motiviert das Studium der Twistor-Konstruktion von nearly ε-Kähler-Mannigfaltigkeiten, welche wir im Detail im dritten Abschnitt dieses Kapitels besprechen. Da die Holonomie des charakteristischen Zusammenhanges für diese Beispiele komplex reduzibel ist, untersuchen wir die Klasse von nearly ε-Kähler-Mannigfaltigkeiten mit dieser zusätzlichen Eigenschaft im vierten Abschnitt. Nachdem wir einige Information in den ersten zwei Teilabschnitten hergeleitet haben, konzentrieren wir uns auf den 6-dimensionalen Fall, da dieser die notwendige geometrische Annahme, welche wir " twistorial type" nennen, für den allgemeinen im vierten Teilabschnitt behandelten Fall illustriert. Im letzten Teilabschnitt kombinieren wir die verschiedenen Informationen, um die inverse Twistor-Konstruktion in Theorem 3.4.18 zu erhalten. Der folgende Abschnitt ist eine Exkursion in unsere Klassifikation flacher pseudo-Riemannscher Liescher Gruppen (Theorem 3.5.1). Im Zusammenspiel mit unseren Klassifikationsergebnissen zu flachen nearly ε-Kähler-Mannigfaltigkeiten im sechsten Abschnitt liefert dies im flachen Falle kompakte Beispiele. Letztere werfen wiederum die Frage nach (nicht flachen, nicht Kählerschen) Ricci-flachen Beispielen auf, die wir mittels einer in den Theoremen 3.7.3 und 3.7.5 gegebenen Korrespondenz zwischen konischen Ricci-flachen para-Kähler-Mannigfaltigkeiten und konischen Ricci-flachen strikten nearly para-Kähler-Mannigfaltigkeiten beantworten. Eine Interpretation dieser Korrespondenz als eine Kegelkonstruktion über para-Sasaki-Einstein-Mannigfaltigkeiten wird in Theorem 3.7.9 bewiesen. Im Kontext von Konstruktionen in einer Kodimension verbleibend verallgemeinern wir im Abschnitt 3.8 als eine Anwendung des obigen äußeren Systems die Evolution von hypo- und nearly hypo-Strukturen in das semi-Riemannsche Setup. Als eine weitere Anwendung des äußeren Systems beobachten wir im Abschnitt 3.9.1, dass Automorphismen gewisser $SU^{\varepsilon}(p,q)$-Strukturen bereits vollständig dadurch bestimmt sind, dass diese eine der beiden stabilen Formen fixieren. Im Abschnitt 3.9.2 erhalten wir die Strukturresultate für $SL(2,\mathbb{R}) \times SL(2,\mathbb{R})$ und erweitern die Ergebnisse für $S^3 \times S^3$, indem wir die nicht-Existenz von (para-)Kähler-Strukturen in indefiniter Signatur zeigen. Die darauf folgenden Unterabschnitte beschäftigen sich mit reell reduzibler Holonomie des charakteristischen Zusammenhanges und dem Bild 3-symmetrischer Räume unter T-Dualität. Im letzten Abschnitt des dritten Kapitels präsentieren wir Resultate zu Lagrange-Untermannigfaltigkeiten in nearly Kähler-Mannigfaltigkeiten. Nachdem wir im ersten Unterabschnitt einige geometrische Informationen zu Lagrange-Untermannigfaltigkeiten in nearly Kähler-Mannigfaltigkeiten abgeleitet haben, zeigen wir im zweiten, dass solche Untermannigfaltigkeiten verschwindende mittlere Krümmung besitzen. Im darauf folgenden Unterabschnitt zeigen wir, dass eine Lagrange-Untermannigfaltigkeit in einer nearly Kähler-Mannigfaltigkeit entsprechend der Zerlegung der nearly Kähler-Mannigfaltigkeit als ein Produkt aus einem Kähler- und einem striktem nearly Kähler-Faktor in ein Produkt zerfällt. Der vierte Unterabschnitt untersucht den Fall, dass der umgebende Raum von der Twistor-Konstruktion herkommt und der letzte Unterabschnitt behandelt Deformationen von Lagrange-Untermannigfaltigkeiten in nearly Kähler-Räumen.

Das vierte Kapitel widmet sich der oben beschriebenen Erweiterung des Hitchin-Flusses. Im ersten Abschnitt geben wir den Lift halbflacher Strukturen zu pseudo-Riemannschen Metriken mit Holonomie $G_2^{(*)}$, besprechen die Vollständigkeit der zugehörigen pseudo-Riemannschen Metriken und setzen kokalibrierte $G_2^{(*)}$-Strukturen zu parallelen Spin(7)- und $\text{Spin}_0(3, 4)$-Strukturen fort. Danach lassen wir nearly Kähler-Metriken zu (nearly) parallelen $G_2^{(*)}$-Strukturen fließen und erhalten auf dem Sinuskegel über diesen Mannigfaltigkeiten mit $G_2^{(*)}$-Struktur Metriken mit Holonomy Spin(7)- und $\text{Spin}_0(3, 4)$. Im dritten Abschnitt geben wir explizite Beispiele auf Nilmannigfaltigkeiten vom Typ $\Gamma \backslash (H_3 \times H_3)$. Hierfür reduzieren wir die Evolution von $\hat{\omega}$ auf einen 4-dimensionalen Unterraum und klassifizieren linksinvariante halbflache Strukturen. Eine der erhaltenen Klassen ist sehr rigide, d.h. entweder flach oder ein Produkt eines para-Hyperkähler symmetrischen Raumes N^4 und eines flachen Faktors (siehe Proposition 4.3.7). Nach diesen Vorbereitungen können wir die Evolutionsgleichungen für alle Klassen invarianter halbflacher Strukturen lösen und Beispiele voller Holonomie G_2 und G_2^* finden (Theorem 4.3.14). Der letzte Abschnitt sammelt die angekündigten Ergebnisse zu speziellen Kähler-Mannigfaltigkeiten.

Chapter 2
Preliminaries

2.1 Stable Forms

In this section we collect some basic facts about stable forms, their orbits and their stabilisers.

Proposition 2.1.1 *Let V be an n-dimensional real or complex vector space. The general linear group $\mathrm{GL}(V)$ has an open orbit in $\Lambda^k V^*$, $0 \leq k \leq \left[\frac{n}{2}\right]$, if and only if $k \leq 2$ or if $k = 3$ and $n = 6, 7$ or 8.*

Proof The representation of $\mathrm{GL}(V)$ on $\Lambda^k V^*$ is irreducible. In the complex case the result thus follows, for instance, from the classification of irreducible complex prehomogeneous vector spaces, [85]. The result in the real case follows from the complex case, since the complexification of the $\mathrm{GL}(n, \mathbb{R})$-module $\Lambda^k \mathbb{R}^{n*}$ is an irreducible $\mathrm{GL}(n, \mathbb{C})$-module. $\qquad\square$

Remark 2.1.2 An open orbit is unique in the complex case, since an orbit which is open in the usual topology is also Zariski-open and Zariski-dense [84, Prop. 2.2]. Over the reals, the number of open orbits is finite by a well-known theorem of Whitney.

Definition 2.1.3 A k-form $\rho \in \Lambda^k V^*$ is called *stable* if its orbit under $\mathrm{GL}(V)$ is open.

Proposition 2.1.4 *Let V be an oriented real vector space and assume that $k \in \{2, n - 2\}$ and n even, or $k \in \{3, n - 3\}$ and $n = 6, 7$ or 8. There is a $\mathrm{GL}(V)^+$-equivariant mapping*

$$\phi : \Lambda^k V^* \to \Lambda^n V^*,$$

homogeneous of degree $\frac{n}{k}$, which assigns a volume form to a stable k-form and which vanishes on non-stable forms. Given a stable k-form ρ, the derivative of ϕ in

© Springer International Publishing AG 2017

L. Schäfer, *Nearly Pseudo-Kähler Manifolds and Related Special Holonomies*,
Lecture Notes in Mathematics 2201, DOI 10.1007/978-3-319-65807-0_2

ρ defines a dual $(n - k)$-form $\hat{\rho} \in \Lambda^{n-k}V^*$ by the property

$$d\phi_\rho(\alpha) = \hat{\rho} \wedge \alpha \quad \text{for all } \alpha \in \Lambda^k V^*. \tag{2.1}$$

The dual form $\hat{\rho}$ is also stable and satisfies[1]

$$(\text{Stab}_{\text{GL}(V)}(\rho))_0 = (\text{Stab}_{\text{GL}(V)}(\hat{\rho}))_0.$$

A stable form, its volume form and its dual are related by the formula

$$\hat{\rho} \wedge \rho = \frac{n}{k}\phi(\rho). \tag{2.2}$$

Proof This result can be viewed as a consequence of the theory of prehomogeneous vector spaces, [84], as follows. Replacing V and $\text{GL}(V)$ by the complexifications $V_{\mathbb{C}}$ and $\text{GL}(V_{\mathbb{C}})$, the situations we are considering correspond to examples 2.3, 2.5, 2.6 and 2.7 of §2, [84]. In all cases, the complement of the open orbit under $\text{GL}(V_{\mathbb{C}})$ is a hypersurface in $\Lambda^k(V_{\mathbb{C}})^*$ defined by a complex irreducible non-degenerate homogeneous polynomial f which is invariant under $\text{GL}(V_{\mathbb{C}})$ up to a non-trivial character $\chi : \text{GL}(V_{\mathbb{C}}) \to \mathbb{C}^\times$.

Due to Proposition 4.1, [84], the polynomial f restricted to $\Lambda^k V^*$ is real-valued and the character χ restricts to $\chi : \text{GL}(V) \to \mathbb{R}^\times$. Moreover, by Proposition 4.5, [84], the complement of the zero set of f in $\Lambda^k V^*$ has a finite number of connected components which are open $\text{GL}(V)$-orbits. Since the only characters of $\text{GL}(V)$ are the powers of the determinant, there is an equivariant mapping from $\Lambda^k V^*$ to $(\Lambda^n V^*)^{\otimes s}$ for some positive integer s. Taking the s-th root, which depends on the choice of an orientation if s is even, we obtain the $\text{GL}^+(V)$-equivariant map ϕ. By construction, a k-form ρ is stable if and only if $\phi(\rho) \neq 0$. The equivariance under scalar matrices implies that the map ϕ is homogeneous of degree $\frac{n}{k}$.

The derivative

$$\Lambda^k V^* \to (\Lambda^k V^*)^* \otimes \Lambda^n V^* \xrightarrow{=} \Lambda^{n-k} V^* \,, \ \rho \mapsto d_\rho \phi \mapsto \hat{\rho}$$

inherits the $\text{GL}^+(V)$-equivariance from ϕ and is an immersion since f is non-degenerate, compare Theorem 2.16, [84]. Therefore, it maps stable forms to stable forms such that the connected components of the stabilisers are identical. Finally, formula (2.2) is just Euler's formula for the homogeneous mapping ϕ.   □

In the following, we discuss stable forms, their volume forms and their duals in the cases which are relevant in this text. In each case, V is a real n-dimensional vector space.

$k = 2, n = 2m$. The orbit of a non-degenerate two-form is open and there is only one open orbit in $\Lambda^2 V^*$. Thus, the stabiliser of a stable two-form ω is isomorphic to $\text{Sp}(2m, \mathbb{R})$. The polynomial invariant is the Pfaffian determinant. We normalise the

---

[1] Here G_0 is the identity component of a Lie group G.

associated equivariant volume form such that it corresponds to the Liouville volume form

$$\phi(\omega) = \frac{1}{m!}\omega^m.$$

Differentiation of the homogeneous polynomial map $\omega \mapsto \phi(\omega)$ yields

$$\hat{\omega} = \frac{1}{(m-1)!}\omega^{m-1}.$$

k = (n − 2), n = 2m. As $\Lambda^{n-2}V^* \cong \Lambda^2 V \otimes \Lambda^n V^*$, there is again only one open orbit. More precisely, an $(n-2)$-form σ is stable if and only if there is a stable two-form ω with $\sigma = \hat{\omega}$ since the mapping $\omega \mapsto \hat{\omega}$ is an equivariant immersion. If m is even, such an ω is unique and we define the volume form $\phi(\sigma) = \phi(\omega)$. If m is odd, we need an orientation on V to uniquely define an associated volume form. We choose the (m-1)-th root ω with positively oriented ω^m and define again $\phi(\sigma) = \phi(\omega)$. In both cases, we find

$$\hat{\sigma} = \frac{1}{m-1}\omega \tag{2.3}$$

with the help of (2.2). The stabiliser of a stable $(n-2)$-form in $GL^+(V)$ is again the real symplectic group.

k = 3, n = 6. Let V be an oriented six-dimensional vector space and let κ denote the canonical isomorphism

$$\kappa : \Lambda^k V^* \cong \Lambda^{6-k} V \otimes \Lambda^6 V^*.$$

Given any three-form ρ, we define $K : V \to V \otimes \Lambda^6 V^*$ by

$$K_\rho(v) = \kappa((v \lrcorner \rho) \wedge \rho)$$

and the quartic invariant

$$\lambda(\rho) = \frac{1}{6}\text{tr}(K_\rho^2) \in (\Lambda^6 V^*)^{\otimes 2}. \tag{2.4}$$

Recall that, for any one-dimensional vector space L, an element $u \in L^{\otimes 2}$ is defined to be positive, $u > 0$, if $u = s \otimes s$ for some $s \in L$ and negative if $-u > 0$. Therefore, the norm of an element $u \in L^{\otimes 2}$ is well-defined and we set

$$\phi(\rho) = \sqrt{|\lambda(\rho)|} \tag{2.5}$$

for the positively oriented square root. If $\phi(\rho) \neq 0$, we furthermore define

$$J_\rho = \frac{1}{\phi(\rho)} K_\rho. \tag{2.6}$$

Proposition 2.1.5 *A three-form ρ on an oriented six-dimensional vector space V with volume form ν is stable if and only $\lambda(\rho) \neq 0$. There are two open orbits.*

One orbit consists of all three-forms ρ satisfying one of the following equivalent properties.

(a) $\lambda(\rho) > 0$.
(b) *There are two uniquely defined real decomposable three-forms α and β such that $\rho = \alpha + \beta$ and $\alpha \wedge \beta > 0$.*
(c) *The stabiliser of ρ in $GL^+(V)$ is $SL(3, \mathbb{R}) \times SL(3, \mathbb{R})$.*
(d) *It holds $\lambda(\rho) \neq 0$ and the endomorphism J_ρ is a para-complex structure on V, i.e. $J^2 = \mathrm{id}_V$ and the eigenspaces for the eigenvalues ± 1 are three-dimensional (see also Sect. 2.2 of the present chapter).*
(e) *There is a basis $\{e_1, \ldots, e_6\}$ of V such that $\nu = e^{123456} > 0$ and*

$$\rho = e^{123} + e^{456}$$

where e^{ijk} is the standard abbreviation for $e^i \wedge e^j \wedge e^k$. In this basis, it holds $\lambda(\rho) = \nu^{\otimes 2}$, $J_\rho e_i = e_i$ for $i \in \{1, 2, 3\}$ and $J_\rho e_i = -e_i$ for $i \in \{4, 5, 6\}$.

The other orbit consists of all three-forms ρ satisfying one of the following equivalent properties.

(a) $\lambda(\rho) < 0$.
(b) *There is a unique complex decomposable three-form α such that $\rho = \alpha + \bar{\alpha}$ and $i(\bar{\alpha} \wedge \alpha) > 0$.*
(c) *The stabiliser of ρ in $GL^+(V)$ is $SL(3, \mathbb{C})$.*
(d) *It holds $\lambda(\rho) \neq 0$ and the endomorphism J_ρ is a complex structure on V.*
(e) *There is a basis $\{e_1, \ldots, e_6\}$ of V such that $\nu = e^{123456} > 0$ and*

$$\rho = e^{135} - e^{146} - e^{236} - e^{245}.$$

In this basis, it holds $\lambda(\rho) = -4\nu^{\otimes 2}$, $J_\rho e_i = -e_{i+1}$ and $J_\rho e_{i+1} = e_i$ for $i \in \{1, 3, 5\}$.

Proof All properties are proved in Section 2 of [75]. The only fact we added is the observation that J_ρ is a para-complex structure if $\lambda(\rho) > 0$ which is obvious in the standard basis. □

It is also possible to introduce a basis describing both orbits simultaneously. Indeed, given a generic stable three-form and an orientation, there is a basis $\{e_1, \ldots, e_6\}$ of V and an $\varepsilon \in \{\pm 1\}$ such that $\nu = e^{123456} > 0$ and

$$\rho_\varepsilon = e^{135} + \varepsilon(e^{146} + e^{236} + e^{245}) \tag{2.7}$$

with $\lambda(\rho) = 4\varepsilon v^{\otimes 2}$. Furthermore, it holds $J_\rho e_i = \varepsilon e_{i+1}, J_\rho e_{i+1} = e_i$ for $i \in \{1, 3, 5\}$ and

$$J_{\rho_\varepsilon}^* \rho_\varepsilon = e^{246} + \varepsilon(e^{235} + e^{145} + e^{136}). \tag{2.8}$$

Analogies between complex and para-complex structures are elaborated in a unified language in Sect. 2.2 of this chapter. In this language, a stable three-form always induces an ε-complex structure J_ρ since $J_\rho^2 = \varepsilon \mathrm{id}$ for the normal form ρ_ε.

Lemma 2.1.6 *The dual of a stable three-form $\rho \in \Lambda^3 V^*$ on an oriented six-dimensional vector space V is*

$$\hat\rho = J_\rho^* \rho. \tag{2.9}$$

Proof We already observed that the connected components of the stabilisers of ρ and $\hat\rho$ have to be identical. Therefore, since the space of real three-forms invariant under $\mathrm{SL}(3, \mathbb{C})$ respectively $\mathrm{SL}(3, \mathbb{R}) \times \mathrm{SL}(3, \mathbb{R})$ is two-dimensional, we can make the ansatz

$$\hat\rho = c_1 \rho + c_2 J_\rho^* \rho$$

with real constants c_1 and c_2. Computing

$$\frac{6}{3}\phi(\rho) \overset{(2.2)}{=} \hat\rho \wedge \rho = c_2 J_\rho^* \rho \wedge \rho \overset{(2.7),(2.8)}{=} 2c_2 \phi(\rho),$$

we find $c_2 = 1$. By

$$d_\rho \phi(J_\rho^* \rho) = \hat\rho \wedge J_\rho^* \rho = c_1 \rho \wedge J_\rho^* \rho = -2c_1 \phi(\rho),$$

the constant c_1 vanishes if the derivative of λ (recall (2.5)) in ρ in direction of $J_\rho^* \rho$ vanishes. However, using the normal form (2.7) again, we compute $\lambda(\rho + tJ_\rho^* \rho) = 4\varepsilon(-\varepsilon + t^2)^2 (e^{123456})^{\otimes 2}$ and the assertion follows. $\qquad\square$

A convenient way to compute the dual of ρ without determining J_ρ is given by the following corollary. In fact, the corollary explicitly shows the equivalence of the two different definitions of $\rho \mapsto \hat\rho$ given in [76] and [75].

Corollary 2.1.7 *If $\lambda(\rho) > 0$ and $\rho = \alpha + \beta$ in terms of decomposables ordered such that $\alpha \wedge \beta > 0$, the dual of ρ satisfies $\hat\rho = \alpha - \beta$.*

If $\lambda(\rho) < 0$ and ρ is the real part of a complex decomposable three-form α such that $i(\bar\alpha \wedge \alpha) > 0$, the dual of ρ is the imaginary part of α. In particular, the complex three-form α is a $(3, 0)$-form w.r.t. J_ρ.

Proof The assertions are easily proved by comparing the claimed formulas for $\hat\rho$ with formula (2.9) in the standard bases given in part (2.1.5) of Proposition 2.1.5. $\qquad\square$

Finally, we note that for a fixed orientation, it holds

$$\hat{\hat\rho} = -\rho \quad \text{and} \quad J_{\hat\rho} = -\varepsilon J_\rho. \tag{2.10}$$

k = 3, n = 7. Given any three-form φ, we define a symmetric bilinear form with values in $\Lambda^7 V^*$ by

$$b_\varphi(v, w) = \frac{1}{6}(v \lrcorner \varphi) \wedge (w \lrcorner \varphi) \wedge \varphi. \tag{2.11}$$

Since the determinant of a scalar-valued bilinear form is an element of $(\Lambda^7 V^*)^{\otimes 2}$, we have $\det b \in (\Lambda^7 V^*)^{\otimes 9}$. If and only if φ is stable, the seven-form

$$\phi(\varphi) = (\det b_\varphi)^{\frac{1}{9}}$$

defines a volume form, independent of an orientation on V, and the scalar-valued symmetric bilinear form

$$g_\varphi = \frac{1}{\phi(\varphi)} b_\varphi$$

is non-degenerate. Notice that $\phi(\varphi) = \sqrt{\det g_\varphi}$ is the metric volume form.

It is known [23, 72] that a stable three-form defines a multiplication "$\cdot$" and a vector cross product "$\times$" on V by the formula

$$\varphi(x, y, z) = g_\varphi(x, y \cdot z) = g_\varphi(x, y \times z), \tag{2.12}$$

such that $(V, \times)$ is isomorphic either to the imaginary octonions $\mathrm{Im}\,\mathbb{O}$ or to the imaginary split-octonions $\mathrm{Im}\,\tilde{\mathbb{O}}$. Thus, there are exactly two open orbits of stable three-forms having isotropy groups

$$\mathrm{Stab}_{\mathrm{GL}(V)}(\varphi) \cong \begin{cases} G_2 \subset \mathrm{SO}(7), & \text{if } g_\varphi \text{ is positive definite}, \\ G_2^* \subset \mathrm{SO}(3, 4), & \text{if } g_\varphi \text{ is of signature (3,4)}. \end{cases} \tag{2.13}$$

There is always a basis $\{e_1, \ldots, e_7\}$ of V such that

$$\varphi = \tau e^{124} + \sum_{i=2}^{7} e^{i\,(i+1)\,(i+3)} \tag{2.14}$$

with $\tau \in \{\pm 1\}$ and indices modulo 7. For $\tau = 1$, the induced metric g_φ is positive definite and the basis is orthonormal such that this basis corresponds to the Cayley basis of $\mathrm{Im}\,\mathbb{O}$. For $\tau = -1$, the metric is of signature (3,4) and the basis is pseudo-orthonormal with e_1, e_2 and e_4 being the three spacelike basis vectors.

The only four-forms having the same stabiliser as φ are the multiples of the Hodge dual $*_{g_\varphi}\varphi$, [23, Propositions 2.1, 2.2]. Since the normal form satisfies $g_\varphi(\varphi, \varphi) = 7$, we have by definition of the Hodge dual $\varphi \wedge *_{g_\varphi}\varphi = 7\phi(\varphi)$ and therefore

$$\hat{\varphi} = \frac{1}{3} *_{g_\varphi} \varphi, \tag{2.15}$$

by comparing with (2.2).

Lemma 2.1.8 *Let φ be a stable three-form in a seven-dimensional vector space V. Let β be a one-form or a two-form. Then $\beta \wedge \varphi = 0$ if and only if $\beta = 0$.*

Proof For the compact case, see also [22]. If β is a one-form, the proof is very easy. If β is a two-form, we choose a basis such that φ is in the normal form (2.14). Let $\beta = \sum_{i<j} b_{i,j} e^{ij}$ and compute

$$
\begin{aligned}
\beta \wedge \varphi = &\,(b_{2,3} - b_{1,6}) \, e^{12356} + (b_{2,3} - b_{4,7}) \, e^{23457} + (b_{1,6} + b_{4,7}) \, e^{14567} \\
&+ (b_{5,7}\tau + b_{1,2}) \, e^{12457} + (b_{3,6} - b_{5,7}) \, e^{34567} + (b_{1,2} - b_{3,6}\tau) \, e^{12346} \\
&- (b_{3,7}\tau + b_{2,4}) \, e^{12347} + (b_{5,6}\tau + b_{2,4}) \, e^{12456} + (b_{3,7} + b_{5,6}) \, e^{13567} \\
&+ (b_{2,5} - b_{4,6}) \, e^{23456} + (b_{4,6} - b_{1,7}) \, e^{13467} - (b_{2,5} + b_{1,7}) \, e^{12357} \\
&+ (b_{4,5} + b_{2,6}) \, e^{24567} - (b_{1,3} + b_{2,6}) \, e^{12367} + (b_{4,5} + b_{1,3}) \, e^{13457} \\
&+ (b_{3,5} + b_{6,7}) \, e^{23567} + (b_{1,4} - b_{3,5}\tau) \, e^{12345} + (b_{6,7}\tau - b_{1,4}) \, e^{12467} \\
&+ (b_{3,4} + b_{1,5}) \, e^{13456} + (b_{2,7} - b_{1,5}) \, e^{12567} + (b_{3,4} - b_{2,7}) \, e^{23467}.
\end{aligned}
$$

This five-form is written as a linear combination of linearly independent forms and each line contains exactly three different coefficients of β. Inspecting the coefficient equations line by line, it is easy to see that all coefficients of β vanish if and only if $\beta \wedge \varphi = 0$. $\qquad\square$

2.1.1 Real Forms of SL(3, $\mathbb{C}$)

By the following proposition, any real form of SL(3, $\mathbb{C}$) can be written as a simultaneous stabiliser of a stable two-form and a stable three-form.

Proposition 2.1.9 *Let V be a six-dimensional real vector space. Let $\omega \in \Lambda^2 V^*$ and $\rho \in \Lambda^3 V^*$ be stable forms which are compatible in the sense that*

$$\omega \wedge \rho = 0. \tag{2.16}$$

Then, we have

$$
\mathrm{Stab}_{\mathrm{GL}(V)}(\rho, \omega) \cong
\begin{cases}
\mathrm{SU}(p,q) \subset \mathrm{SO}(2p, 2q), & p + q = 3, \quad \text{if } \lambda(\rho) < 0, \\
\mathrm{SL}(3, \mathbb{R}) \subset \mathrm{SO}(3,3), & \text{if } \lambda(\rho) > 0,
\end{cases}
$$

where $SL(3, \mathbb{R})$ *is embedded in* $SO(3, 3)$ *such that it acts by the standard representation and its dual, respectively, on the maximally isotropic* ± 1-*eigenspaces of the para-complex structure* J_ρ *induced by* ρ.

Proof Let V be oriented by $\phi(\omega) = \frac{1}{6}\omega^3$ and let J_ρ be the unique (para-) complex structure (2.6) associated to the three-form ρ and this orientation. By $\varepsilon \in \{\pm 1\}$, we denote the sign of $\lambda(\rho)$, that is $J_\rho^2 = \varepsilon id_V$. In the basis in which ρ is in the normal form (2.7), it is easy to verify that $\omega \wedge \rho = 0$ is equivalent to the skew-symmetry of J_ρ with respect to ω. Equivalently, the pseudo-Euclidean metric

$$g = g_{(\omega,\rho)} = \varepsilon\, \omega(\cdot, J_\rho \cdot), \qquad (2.17)$$

induced by ω and ρ, is compatible with J_ρ in the sense that $g(J_\rho \cdot, J_\rho \cdot) = -\varepsilon g(\cdot, \cdot)$. The stabiliser of the set of tensors $(\omega, J_\rho, g, \rho, J_\rho^* \rho)$ satisfying this compatibility condition is well-known to be $SU(p, q)$ respectively $SL(3, \mathbb{R})$. $\square$

We call a compatible pair of stable forms $(\omega, \rho) \in \Lambda^2 V^* \times \Lambda^3 V^*$ normalised if

$$\phi(\rho) = 2\phi(\omega) \qquad \Longleftrightarrow \qquad J_\rho^* \rho \wedge \rho = \frac{2}{3}\omega^3. \qquad (2.18)$$

Moreover, a stable three-form ρ is non-degenerate in the sense that for $v \in V$ one has

$$v \lrcorner \rho = 0 \;\Rightarrow\; v = 0. \qquad (2.19)$$

Remark 2.1.10 By our conventions, the metric (2.17) induced by a normalised, compatible pair is either of signature $(6, 0)$ or $(2, 4)$ or $(3, 3)$, where the first number denotes the number of spacelike directions. We emphasise that our conventions are such that

$$\omega = g(\cdot, J_\rho \cdot).$$

This sign choice turned out to be necessary in order to achieve that $\phi(\rho)$ is indeed a positive multiple of $\phi(\omega)$ in the positive definite case.

Sometimes it is convenient to have a unified adapted basis. For a compatible and normalised pair (ω, ρ), there is always a pseudo-orthonormal basis $\{e_1, \ldots, e_6\}$ of V with dual basis $\{e^1, \ldots, e^6\}$ such that $\rho = \rho_\varepsilon$ is in the normal form (2.7) and

$$\omega = \tau(e^{12} + e^{34}) + e^{56} \qquad (2.20)$$

for $(\varepsilon, \tau) \in \{(-1, 1), (-1, -1), (1, 1)\}$. The signature of the induced metric with respect to this basis is

$$(\tau, -\varepsilon\tau, \tau, -\varepsilon\tau, 1, -\varepsilon) = \begin{cases} (+, +, +, +, +, +) \text{ for } \varepsilon = -1 \text{ and } \tau = 1, \\ (-, -, -, -, +, +) \text{ for } \varepsilon = -1 \text{ and } \tau = -1, \quad (2.21) \\ (+, -, +, -, +, -) \text{ for } \varepsilon = 1 \text{ and } \tau = 1, \end{cases}$$

and we have

$$\mathrm{Stab}_{GL(6,\mathbb{R})}(\omega,\rho) \cong \begin{cases} SU(3) \subset SO(6) & \text{for } \varepsilon = -1 \text{ and } \tau = 1, \\ SU(1,2) \subset SO(2,4) & \text{for } \varepsilon = -1 \text{ and } \tau = -1, \\ SL(3,\mathbb{R}) \subset SO(3,3) & \text{for } \varepsilon = 1. \end{cases}$$

For instance, the following observation is easily verified using the unified basis.

Lemma 2.1.11 *Let* (ω,ρ) *be a compatible and normalised pair of stable forms on a six-dimensional vector space. Then, the volume form* $\phi(\omega)$ *is in fact a metric volume form w.r.t. to the induced metric* $g = g_{(\omega,\rho)}$ *and the corresponding Hodge dual of* ω *and* ρ *is*

$$*_g \omega = -\varepsilon\hat{\omega}, \qquad *_g \rho = -\hat{\rho} \qquad (2.22)$$

2.1.2 Relation Between Real Forms of $SL(3,\mathbb{C})$ and $G_2^{\mathbb{C}}$

The relation between stable forms in dimension 6 and 7 corresponding to the embedding $SU(3) \subset G_2$ is well-known. We extend this relation by including also the embeddings $SU(1,2) \subset G_2^*$ and $SL(3,\mathbb{R}) \subset G_2^*$ as follows.

Proposition 2.1.12 *Let* $V = W \oplus L$ *be a seven-dimensional vector space decomposed as a direct sum of a six-dimensional subspace* W *and a line* L. *Let* α *be a non-trivial one-form in the annihilator* W^0 *of* W *and* $(\omega,\rho) \in \Lambda^2 L^0 \times \Lambda^3 L^0$ *a compatible and normalised pair of stable forms inducing the scalar product* $h = h_{(\omega,\rho)}$ *given in (2.17). Then, the three-form* $\varphi \in \Lambda^3 V^*$ *defined by*

$$\varphi = \omega \wedge \alpha + \rho \qquad (2.23)$$

is stable and induces the scalar product

$$g_\varphi = h - \varepsilon\alpha \cdot \alpha, \qquad (2.24)$$

where ε *denotes the sign of* $\lambda(\rho)$ *such that* $J_\rho^2 = \varepsilon\mathrm{id}$. *The stabiliser of* φ *in* $GL(V)$ *is*

$$\mathrm{Stab}_{GL(V)}(\varphi) \cong \begin{cases} G_2 & \text{for } \varepsilon = -1 \text{ and positive definite } h, \\ G_2^* & \text{otherwise.} \end{cases}$$

Proof We choose a basis $\{e_1, \dots, e_6\}$ of L^0 such that ω and ρ are in the generic normal forms (2.7) and (2.20). With $e^7 = \alpha$, we have

$$\varphi = \tau(e^{127} + e^{347}) + e^{567} + e^{135} + \varepsilon(e^{146} + e^{236} + e^{245}). \qquad (2.25)$$

The induced bilinear form (2.11) turns out to be

$$b_\varphi(v, w) = (-\varepsilon\tau v^1 w^1 + \tau v^2 w^2 - \varepsilon\tau v^3 w^3 + \tau v^4 w^4 - \varepsilon v^5 w^5 + v^6 w^6 + v^7 w^7) e^{1234567}$$

for $v = \sum v^i e_i$ and $w = \sum w^i e_i$. Hence, the three-form φ is stable for all signs of ε and τ and its associated volume form is

$$\phi(\varphi) = (\det b_\varphi)^{\frac{1}{9}} = -\varepsilon\, e^{1234567}.$$

The formula (2.24) for the metric g_φ induced by φ follows, since the basis $\{e_1, \ldots, e_7\}$ of V is pseudo-orthonormal with respect to this metric of signature

$$(\tau, -\varepsilon\tau, \tau, -\varepsilon\tau, 1, -\varepsilon, -\varepsilon) = \begin{cases} (+, +, +, +, +, +, +) \text{ for } \varepsilon = -1 \text{ and } \tau = 1, \\ (-, -, -, -, +, +, +) \text{ for } \varepsilon = -1 \text{ and } \tau = -1, \quad (2.26) \\ (+, -, +, -, +, -, -) \text{ for } \varepsilon = 1 \text{ and } \tau = 1. \end{cases}$$

The assertion on the stabilisers now follows from (2.13).   □

Lemma 2.1.13 *Under the assumptions of the previous proposition, the dual four-form of the stable three-form φ is*

$$3\hat{\varphi} = *_\varphi\varphi = -\varepsilon\,(\alpha \wedge \hat{\rho} + \hat{\omega}) = \varepsilon\alpha \wedge *_h\rho + *_h\omega, \qquad (2.27)$$

where $*_\varphi$ denotes the Hodge dual with respect to the metric g_φ and the orientation induced by $\phi(\varphi)$.

Proof In the basis of the previous proof, the Hodge dual of φ is

$$*_\varphi\varphi = -\varepsilon\tau(e^{3456} + e^{1256}) - \varepsilon e^{1234} + \varepsilon\, e^{2467} + e^{2357} + e^{1457} + e^{1367}.$$

The second equality follows when comparing this expression with $\varepsilon(e^7 \wedge \hat{\rho} + \frac{1}{2}\omega^2)$ in this basis using (2.8) and (2.20). The first and the third equality are just the formulas (2.15) and (2.22), respectively.   □

The inverse process is given by the following construction.

Proposition 2.1.14 *Let V be a seven-dimensional real vector space and $\varphi \in \Lambda^3 V^*$ a stable three-form which induces the metric g_φ on V. Moreover, let $n \in V$ be a unit vector with $g_\varphi(n, n) = -\varepsilon \in \{\pm 1\}$ and let $W = n^\perp$ denote the orthogonal complement of $\mathbb{R}\cdot n$. Then, the pair $(\omega, \rho) \in \Lambda^2 W^* \times \Lambda^3 W^*$ defined by*

$$\omega = (n \lrcorner \varphi)_{|W}, \qquad \rho = \varphi_{|W}, \qquad (2.28)$$

is a pair of compatible normalised stable forms. The metric $h = h_{(\omega, \rho)}$ induced by this pair on W satisfies $h = (g_\varphi)_{|W}$ and the stabiliser is

$$\mathrm{Stab}_{\mathrm{GL}(W)}(\omega, \rho) \cong \begin{cases} \mathrm{SU}(3), & \textit{if } g_\varphi \textit{ is positive definite,} \\ \mathrm{SU}(1, 2), & \textit{if } g_\varphi \textit{ is indefinite and } \varepsilon = -1, \\ \mathrm{SL}(3, \mathbb{R}), & \textit{if } \varepsilon = 1. \end{cases}$$

When (V, φ) is identified with the imaginary octonions, respectively, the imaginary split-octonions, by (2.12), the ε-complex structure induced by ρ is given by

$$J_\rho v = -n \cdot v = -n \times v \qquad\qquad \textit{for } v \in V. \qquad (2.29)$$

Proof Due to the stability of φ, we can always choose a basis $\{e_1, \ldots, e_7\}$ of V with $n = e_7$ such that φ is given by (2.25) where $\varepsilon = -g_\varphi(n, n)$ and $\tau \in \{\pm 1\}$ depends on the signature of g_φ. As this basis is pseudo-orthonormal with signature given by (2.26), the vector n has indeed the right scalar square and $\{e_1, \ldots, e_6\}$ is a pseudo-orthonormal basis of the complement $W = n^\perp$. Since the pair (ω, ρ) defined by (2.28) is now exactly in the generic normal form given by (2.7) and (2.20), it is stable, compatible and normalised and the induced endomorphism J_ρ is an ε-complex structure. The identity $h = (g_\varphi)_{|W}$ for the induced metric $h_{(\omega, \rho)}$ follows from comparing the signatures (2.26) and (2.21) and the assertion for the stabilisers is an immediate consequence. Finally, the formula for the induced ε-complex structure J_ρ is another consequence of $g = (g_\varphi)_{|W}$ since we have

$$g_\varphi(x, n \times y) \overset{(2.12)}{=} \varphi(x, n, y) = -\omega(x, y) = -h(x, J_\rho y)$$

for all $x, y \in W$.   □

Notice that, for a fixed metric h of signature $(2, 4)$ or $(3, 3)$, the compatible and normalised pairs (ω, ρ) of stable forms inducing this metric are parametrised by the homogeneous spaces $\mathrm{SO}(2, 4)/\mathrm{SU}(1, 2)$ and $\mathrm{SO}(3, 3)/\mathrm{SL}(3, \mathbb{R})$, respectively. Thus, the mapping $(\omega, \rho) \mapsto \varphi$ defined by formula (2.23) yields isomorphisms

$$\frac{\mathrm{SO}(2, 4)}{\mathrm{SU}(1, 2)} \cong \frac{\mathrm{SO}(3, 4)}{\mathrm{G}_2^*}, \qquad \frac{\mathrm{SO}(3, 3)}{\mathrm{SL}(3, \mathbb{R})} \cong \frac{\mathrm{SO}(3, 4)}{\mathrm{G}_2^*},$$

since the metric h completely determines the metric g_φ by the formula (2.24).

2.1.3 Relation Between Real Forms of $\mathrm{G}_2^{\mathbb{C}}$ and $\mathrm{Spin}(7, \mathbb{C})$

It is possible to extend this construction to dimension 8 as follows. Starting with a stable three-form φ on a seven-dimensional vector space V, we can consider the four-form

$$\Phi = e^8 \wedge \varphi + *_\varphi \varphi. \tag{2.30}$$

on the eight-dimensional space $V \oplus \mathbb{R}e_8$. Although the four-form Φ is not stable, it is shown in [23] that it induces the metric

$$g_\Phi = g_\varphi + (e^8)^2 \tag{2.31}$$

on $V \oplus \mathbb{R}e_8$ and that its stabiliser is

$$\operatorname{Stab}_{\mathrm{GL}(V \oplus \mathbb{R}e_8)}(\Phi) \cong \begin{cases} \operatorname{Spin}(7) \subset \mathrm{SO}(8), & \text{if } g_\varphi \text{ is positive definite,} \\ \operatorname{Spin}_0(3,4) \subset \mathrm{SO}(4,4), & \text{if } g_\varphi \text{ is indefinite.} \end{cases}$$

The index "0" denotes, as usual, the connected component of the identity. Starting conversely with a four-form Φ on $V \oplus \mathbb{R}e_8$ such that its stabiliser in $\mathrm{GL}(V \oplus \mathbb{R}e_8)$ is isomorphic to $\operatorname{Spin}(7)$ or $\operatorname{Spin}_0(3,4)$, the process can be reversed by setting $\varphi = e_8 \lrcorner \Phi$. As before, the metric induced by Φ on $V \oplus \mathbb{R}e_8$ is determined by the metric g_φ induced by φ on V. Thus, the indefinite analogue of the well-known isomorphisms

$$\mathbb{RP}^7 \cong \frac{\mathrm{SO}(6)}{\mathrm{SU}(3)} \cong \frac{\mathrm{SO}(7)}{\mathrm{G}_2} \cong \frac{\mathrm{SO}(8)}{\operatorname{Spin}(7)}$$

is given by

$$\frac{\mathrm{SO}(2,4)}{\mathrm{SU}(1,2)} \cong \frac{\mathrm{SO}(3,3)}{\mathrm{SL}(3,\mathbb{R})} \cong \frac{\mathrm{SO}(3,4)}{\mathrm{G}_2^*} \cong \frac{\mathrm{SO}(4,4)}{\operatorname{Spin}_0(3,4)}. \tag{2.32}$$

2.2 Almost Pseudo-Hermitian and Almost Para-Hermitian Geometry

We recall that an almost para-complex structure on a $2m$-dimensional manifold M is an endomorphism field squaring to the identity such that both eigendistributions (for the eigenvalues ± 1) are m-dimensional. An almost para-Hermitian structure consists of a neutral metric and an antiorthogonal almost para-complex structure. For a survey on para-complex geometry we refer to [9] or [47].

In the following, we introduce a unified language describing almost pseudo-Hermitian and almost para-Hermitian geometry simultaneously. The philosophy is to put an "ε" in front of all notions which is to be replaced by "para" for $\varepsilon = 1$ and is to be replaced by "pseudo" or to be omitted for $\varepsilon = -1$. From now on, we always suppose $\varepsilon \in \{\pm 1\}$. Let us note, that there exist other names in the literature, like for example *double numbers* or *hypercomplex numbers*. To begin with, we consider the ε-complex numbers $\mathbb{C}_\varepsilon = \{x + i_\varepsilon y, x, y \in \mathbb{R}\}$ with $i_\varepsilon^2 = \varepsilon$. For the para-complex numbers, $\varepsilon = 1$, there are obvious analogues of conjugation, real and imaginary parts and the square of the (not necessarily positive) absolute value given by $|z|^2 = z\bar{z}$.

Moreover, let V be a real vector space of even dimension $n = 2m$. We call an endomorphism J^ε an ε-complex structure if $J^{\varepsilon 2} = \varepsilon \mathrm{id}_V$ and if additionally, for $\varepsilon = 1$, the ± 1-eigenspaces $V^\pm$ are m-dimensional. If we consider only one of the two cases, we write J for $\varepsilon = -1$ and P or τ for $\varepsilon = 1$. An ε-Hermitian structure is an ε-complex structure J^ε together with a pseudo-Euclidean scalar-product g which is ε-Hermitian in the sense that it holds

$$g(J^\varepsilon\cdot, J^\varepsilon\cdot) = -\varepsilon g(\cdot, \cdot).$$

We denote the stabiliser in $\mathrm{GL}(V)$ of an ε-Hermitian structure as the ε-unitary group

$$\mathrm{U}(J^\varepsilon, g) = \mathrm{U}^\varepsilon(p, q) = \{L \in \mathrm{GL}(V) \mid [L, J^\varepsilon] = 0, L^* g = g\}$$

$$\cong \begin{cases} \mathrm{U}(p, q), \ p + q = m, & \text{for } \varepsilon = -1, \\ \mathrm{GL}(m, \mathbb{R}), & \text{for } \varepsilon = 1. \end{cases}$$

Here, the pair $(2p, 2q)$ is the signature[2] of the metric for $\varepsilon = -1$. For $\varepsilon = 1$, the group $\mathrm{GL}(m, \mathbb{R})$ acts reducibly such that $V = V^+ \oplus V^-$ and the signature is always (m, m). In other words, there exists a basis called isotropic or null-coordinates of V such that P and g are represented by the matrices

$$g_{can} = g_0 = \begin{pmatrix} 0 & Id \\ Id & 0 \end{pmatrix} \text{ and } P_{can} = P_0 = \begin{pmatrix} Id & 0 \\ 0 & -Id \end{pmatrix} \tag{2.33}$$

and one computes that

$$\mathrm{U}(P, g) \cong \left\{ A \in \mathrm{GL}(\mathbb{R}^m) \left| \begin{pmatrix} A & 0 \\ 0 & (A^t)^{-1} \end{pmatrix} \right. \right\} \cong \mathrm{GL}(\mathbb{R}^m).$$

Later we consider the case, where the para-Kähler manifold is Ricci-flat, then the restricted holonomy group lies in the special para-unitary group $SU(P, g)$ which in the above model is

$$SU(P, g) \cong \left\{ A \in SL(\mathbb{R}^m) \left| \begin{pmatrix} A & 0 \\ 0 & (A^t)^{-1} \end{pmatrix} \right. \right\} \cong SL(\mathbb{R}^m), \tag{2.34}$$

see also Proposition 2.1.9.

---

[2]Please note that in our convention $2p$ refers to the negative directions.

An almost ε-Hermitian manifold is a manifold M of dimension $n = 2m$ endowed with a $U^{\varepsilon}(p, q)$-structure or, equivalently, with an almost ε-Hermitian structure which consists of an almost ε-complex structure J^{ε} and an ε-Hermitian metric g. The non-degenerate two-form $\omega := g(\cdot, J^{\varepsilon}\cdot)$ is called fundamental two-form.

Given an almost ε-Hermitian structure $(g, J^{\varepsilon}, \omega)$, there exist pseudo-orthonormal local frames $\{e_1, \ldots, e_{2m}\}$ such that $Je_i = e_{i+m}$ for $i = 1, \ldots, m$ and $\omega = \varepsilon \sum_{i=1}^{m} \sigma_i e^{i(i+m)}$, where $\sigma_i := g(e_i, e_i)$ for $i = 1, \ldots, m$. Upper indices always denote dual (not metric dual) one-forms and e^{ij} stands for $e^i \wedge e^j$. We call such a frame ε-unitary. If $m \geq 3$, we can always achieve $\sigma_1 = \sigma_2$ by reordering the basis vectors.

For both values of ε, the ε-complexification $TM \otimes \mathbb{C}_{\varepsilon}$ of the tangent bundle decomposes into the $\pm i_{\varepsilon}$-eigenbundles $TM^{1,0}$ and $TM^{0,1}$. This induces the well-known bi-grading of $\mathbb{C}_{\varepsilon}$-valued exterior forms

$$\Omega^{r,s} = \Omega^{r,s}(M) = \Gamma(\Lambda^{r,s}) = \Gamma(\Lambda^r(TM^{1,0})^* \otimes \Lambda^s(TM^{0,1})^*).$$

If X is a vector field on M, we use the notation

$$X^{1,0} = \frac{1}{2}(X + i_{\varepsilon}\varepsilon J^{\varepsilon}X) \in \Gamma(TM^{1,0}), \ X^{0,1} = \frac{1}{2}(X - i_{\varepsilon}\varepsilon J^{\varepsilon}X) \in \Gamma(TM^{0,1}),$$

for the real isomorphisms from TM to $TM^{1,0}$ respectively $TM^{0,1}$. As usual in almost Hermitian geometry [104], we define the bundles $[\![\Lambda^{r,s}]\!]$ for $r \neq s$ and $[\Lambda^{r,r}]$ by the property

$$[\![\Lambda^{r,s}]\!] \otimes \mathbb{C}_{\varepsilon} = [\![\Lambda^{r,s}]\!] \oplus i_{\varepsilon}[\![\Lambda^{r,s}]\!] = \Lambda^{r,s} \oplus \Lambda^{s,r},$$

$$[\Lambda^{r,r}] \otimes \mathbb{C}_{\varepsilon} = [\Lambda^{r,r}] \oplus i_{\varepsilon}[\Lambda^{r,r}] = \Lambda^{r,r}.$$

The sections of these bundles are denoted as real forms of type $(r, s) + (s, r)$ respectively of type (r, r) and the spaces of sections by $[\![\Omega^{r,s}]\!]$ respectively by $[\Omega^{r,r}]$. For instance, it holds

$$[\Omega^{1,1}] = \{\alpha \in \Omega^2 M \,|\, \alpha(X, Y) = -\varepsilon\alpha(J^{\varepsilon}X, J^{\varepsilon}Y)\},$$

such that the fundamental form is of type $(1, 1)$ and similarly

$$[\![\Omega^{3,0}]\!] = \{\alpha \in \Omega^3 M \,|\, \alpha(X, Y, Z) = \varepsilon\alpha(X, J^{\varepsilon}Y, J^{\varepsilon}Z)\}. \tag{2.35}$$

Only in the para-complex case, $\varepsilon = 1$, there is a decomposition of the real tangent bundle $TM = \mathcal{V} \oplus \mathcal{H}$ into the ± 1-eigenbundles of J^{ε} which also induces a bi-grading of real differential forms. It is also straightforward to show that

$$[\![\Lambda^{3,0}]\!] \cong \Lambda^3\mathcal{V}^* \oplus \Lambda^3\mathcal{H}^*, \tag{2.36}$$

when considering the characterisation (2.35).

Returning to analogies, we recall that the Nijenhuis tensor of the almost ε-complex structure J^ε satisfies

$$N(X,Y) = -\varepsilon[X,Y] - [J^\varepsilon X, J^\varepsilon Y] + J^\varepsilon[J^\varepsilon X, Y] + J^\varepsilon[X, J^\varepsilon Y]$$
$$= -(\nabla_{J^\varepsilon X} J^\varepsilon)\, Y + (\nabla_{J^\varepsilon Y} J^\varepsilon)\, X + J^\varepsilon(\nabla_X J^\varepsilon)\, Y - J^\varepsilon(\nabla_Y J^\varepsilon)\, X \quad (2.37)$$

for real vector fields X, Y, Z and for any torsion-free connection ∇ on M. For both values of ε, it is well-known that the Nijenhuis tensor is the obstruction to the integrability of the almost ε-complex structure. Since integrability for complex structures is well-known, let us just note, that a para-complex structure on M is called integrable if the distributions $T^\pm M$ are integrable distributions.

In the following, let ∇ always denote the Levi-Civita connection of the metric g of an almost ε-Hermitian manifold. Differentiating the almost ε-complex structure, its square and the fundamental two-form yields for both values of ε the formulas

$$(\nabla_X J^\varepsilon)\, J^\varepsilon Y = -J^\varepsilon(\nabla_X J^\varepsilon)\, Y,$$
$$g((\nabla_X J^\varepsilon)Y, Z) = -(\nabla_X \omega)(Y, Z) \quad (2.38)$$

for all vector fields X, Y, Z. Using these formulas, it is easy to show that for any almost ε-Hermitian manifold, the tensor $\mathcal{A}$ defined by

$$\mathcal{A}(X,Y,Z) = g((\nabla_X J^\varepsilon)Y, Z) = -(\nabla_X \omega)(Y, Z) \quad (2.39)$$

has the symmetries

$$\mathcal{A}(X,Y,Z) = -\mathcal{A}(X,Z,Y), \quad (2.40)$$
$$\mathcal{A}(X,Y,Z) = \varepsilon \mathcal{A}(X, J^\varepsilon Y, J^\varepsilon Z) \quad (2.41)$$

for all vector fields X, Y, Z.

The decomposition of the $U^\varepsilon(p, q)$-representation space of tensors with the same symmetries as $\mathcal{A}$ into irreducible components leads to a classification of almost ε-Hermitian manifolds which is classical for $U(m)$ [68] and is called the Gray-Hervella list. The para-complex case for the group $GL(m, \mathbb{R})$ is completely worked out in [59]. In [86], the Gray-Hervella classes are generalised to almost ε-Hermitian structures, which are denoted by generalised almost Hermitian or $\mathcal{GAH}$ structures there. Analogues of all 16 Gray-Hervella classes are established. These are invariant under the respective group action, but obviously not irreducible for the para-Hermitian case when compared to the decomposition in [59].

Finally, we mention the useful formula

$$2(\nabla_X \omega)(Y, Z) = d\omega(X, Y, Z) + \varepsilon d\omega(X, J^\varepsilon Y, J^\varepsilon Z) + \varepsilon g(N(Y, Z), J^\varepsilon X) \quad (2.42)$$

holding true for all vector fields X, Y, Z on any almost ε-Hermitian manifold. Proofs can be found in Chapter IX of [87] for $\varepsilon = -1$ and [80] for $\varepsilon = 1$.

2.3   Linear Algebra of Three-Forms in Dimension 8 and 10

In this subsection we shall freely identify the real vector space $V := \mathbb{C}^{k,l}_{-1} = \mathbb{C}^{k,l} = \mathbb{R}^{2k,2l} = \mathbb{R}^{2n}$ and $V := \mathbb{C}^n_1 = \mathbb{R}^{n,n} = \mathbb{R}^{2n}$ with its dual V^* by means of the pseudo-Euclidian scalar product $g = g_{can}$ and we denote the canonical ε-complex structure on V by J^ε_{can}.

Definition 2.3.1 The support of $\eta \in \Lambda^3 V$ is defined by

$$\Sigma_\eta := \text{span}\{\eta_X Y \mid X, Y \in V\} \subset V. \tag{2.43}$$

The name support is motivated by the following Lemma of [41].

Lemma 2.3.2 *Let $\eta \in \Lambda^3 V$, then one has $\eta \in \Lambda^3 \Sigma_\eta$.*

Proof We take a complement $W \subset V$ of $L = \Sigma_\eta$. The decomposition

$$\Lambda^3 V = \bigoplus_{p+q=3} \Lambda^p L \wedge \Lambda^q W$$

induces a decomposition

$$\eta = \sum_{p+q=3} \eta^{[p,q]}.$$

Taking $X, Y \in L^0$ yields[3] $L \ni \eta_X Y = \eta^{[0,3]}_X Y + \eta^{[1,2]}_X Y$. Now since $\eta^{[0,3]}_X Y \in W$ and $\eta^{[1,2]}_X Y \in L$, we get $\eta^{[0,3]} = 0$. Further the choice $X \in L^0$ and $Y \in W^0$ yields $\eta^{[1,2]} = 0$ and then the choice $X, Y \in W^0$ yields $\eta^{[2,1]} = 0$. This shows $\eta = \eta^{[3,0]}$.
$\square$

In Chap. 3 of the present text we are interested in the tensor field $g_p((\nabla_X J)Y, Z)$ for $p \in M$ and $X, Y, Z \in T_p M$ on a nearly ε-Kähler manifold (M, J, g), which is shown to be a real form of type $(3, 0) + (0, 3)$. The type condition implies that Σ_η is a J^ε-invariant subspace. For example, it follows that the complex dimension of the support of a non-zero such form is at least three. In Sect. 3.6 of Chap. 3 the classification of Levi-Civita flat nearly ε-Kähler manifolds is related to the existence of real three-forms of type $(3, 0) + (0, 3)$ with isotropic support, i.e. such that Σ_η is an isotropic subspace. This is our reason for the interest in the subsequent Lemmata [41, 43].

---

[3]As above 0 denotes the annihilator.

Lemma 2.3.3 *One has*

$$\llbracket \Omega^{3,0} \rrbracket = \{\eta \in \Lambda^3 V \mid \eta(\cdot, J^\varepsilon_{can}\cdot, J^\varepsilon_{can}\cdot) = \varepsilon\eta(\cdot,\cdot,\cdot)\}$$

$$= \{\eta \in \Lambda^3 V \mid \{\eta_X, J^\varepsilon_{can}\} = 0, \forall X \in V\}. \tag{2.44}$$

Later in this text we need to consider three-forms which satisfy the condition

$$\eta_X \circ \eta_Y = 0, \ \forall X, Y \in V, \tag{2.45}$$

where η_X for some fixed $X \in V$ is seen as an endomorphism of V. Therefore we show the next Lemma.

Lemma 2.3.4 *The support Σ_η of $\eta \in \Lambda^3 V$ is an isotropic subspace if and only if η satisfies the condition (2.45). If η is of real type $(3,0) + (0,3)$, then Σ_η is J^ε_{can}-invariant.*

Proof First the isotropy of Σ_η is equivalent to $g(\eta_X Y, \eta_Z W) = 0$ for all $X, Y, Z, W \in V$.
From

$$g(\eta_X Y, Z) = \eta(X, Y, Z) = -\eta(X, Z, Y) = -g(\eta_X Z, Y) = -g(Y, \eta_X Z), \ \forall X, Y, Z$$

we get

$$g(\eta_X Y, \eta_Z W) = -g(\eta_Z \eta_X Y, W).$$

This equation shows $\eta_X \circ \eta_Y = 0$, for all $X, Y \in V$, if and only if its support Σ_η is isotropic.

The last assertion follows from

$$J^\varepsilon_{can} \Sigma_\eta = \text{span}\{J^\varepsilon_{can} \eta_X Y \mid X, Y \in V\} \overset{Eq.\,(2.44)}{=} \text{span}\{-\eta_X J^\varepsilon_{can} Y \mid X, Y \in V\} = \Sigma_\eta,$$

where we used that η is of real type $(3,0) + (0,3)$, if and only if η_X anti-commutes with J^ε_{can}. $\qquad\square$

We define the kernel of a three-form $\eta \in \Lambda^3 V^*$ by $\mathcal{K} = \mathcal{K}_\eta = \ker(X \mapsto X \lrcorner \eta)$.

Lemma 2.3.5 (Lemma 3.1 of [108]) *One has $\mathcal{K} = \Sigma_\eta^\perp$ and $\Sigma_\eta = \mathcal{K}^\perp$.*

Lemma 2.3.6 (Lemma 3.2 of [108]) *Let $(V, J, \langle\cdot,\cdot\rangle)$ be a pseudo-Hermitian vector space with $\dim_\mathbb{R}(V) = 8$ then a real three-form η of type $(3,0) + (0,3)$ and of non-vanishing length has a (complex) one dimensional kernel $\mathcal{K}_\eta$, which admits an orthogonal complement $(\mathcal{K}_\eta)^\perp$. Moreover, one has $\Sigma_\eta = (\mathcal{K}_\eta)^\perp$.*

Remark 2.3.7 Let us observe, that, if η has length zero, one can replace the orthogonal complement by the null-space and obtain an analogous statement as in the last proposition.

Lemma 2.3.8 (Lemma 3.4 of [108]) *Let* $(V, J, \langle \cdot, \cdot \rangle)$ *be a pseudo-Hermitian vector space with* $\dim_{\mathbb{R}}(V) = 10$ *then a real three-form* η *of type* $(3, 0) + (0, 3)$ *and of non-vanishing length is of the following possible types:*

(i) *There exists an orthonormal real basis* $\{f_i\}_{i=1}^{10} = \{e_1, Je_1, \ldots, e_5, Je_5\}$ *and real numbers* α, β *such that*

$$\eta(e_1, e_2, e_3) = \alpha \neq 0; \quad \eta(e_4, e_5, e_1) = \beta \qquad (2.46)$$

and $\eta(f_i, f_j, f_k) = 0$ *for the cases which are not obtained from (2.46) by skew-symmetry and type relations.*

(ii) *There exists an orthonormal real basis* $\{f_i\}_{i=1}^{10} = \{e_1, Je_1, \ldots, e_5, Je_5\}$ *and real numbers* α, β *such that*

$$\eta(e_1, e_2, e_3) = \alpha \neq 0; \quad \eta(e_4, e_5, e_1 + e_3) = \beta \text{ with } \langle e_1, e_1 \rangle = -\langle e_3, e_3 \rangle$$
$$(2.47)$$

and $\eta(f_i, f_j, f_k) = 0$ *for the cases which are not obtained from (2.47) by skew-symmetry and type relations.*

2.4   Structure Reduction of Almost ε-Hermitian Six-Manifolds

Let $(V, g, J^\varepsilon, \omega)$ be a $2m$-dimensional ε-Hermitian vector space and $\Psi = \psi^+ + i_\varepsilon \psi^-$ be an $(m, 0)$-form of *non-zero* length. We define the special ε-unitary group $\mathrm{SU}^\varepsilon(p, q)$ as the stabiliser of Ψ in the ε-unitary group $\mathrm{U}^\varepsilon(p, q)$ such that

$$\mathrm{SU}^\varepsilon(p, q) = \mathrm{Stab}_{\mathrm{GL}(V)}(g, J^\varepsilon, \Psi) \cong \begin{cases} \mathrm{SU}(p, q)\,, p + q = m, & \text{for } \varepsilon = -1, \\ \mathrm{SL}(m, \mathbb{R}), & \text{for } \varepsilon = 1, \end{cases}$$

where $\mathrm{SL}(m, \mathbb{R})$ acts reducibly such that $V = V^+ \oplus V^-$.

With this notation, an $\mathrm{SU}^\varepsilon(p, q)$-**structure** on a manifold M^{2m} is an almost ε-Hermitian structure $(g, J^\varepsilon, \omega)$ together with a global $(m, 0)$-form Ψ of non-zero constant length. Locally, there exists an ε-unitary frame $\{e_1, \ldots, e_m, e_{m+1} = Je_1, \ldots, e_{2m} = Je_m\}$ which is adapted to the $\mathrm{SU}^\varepsilon(p, q)$-reduction Ψ in the sense that

$$\Psi = a(e^1 + i_\varepsilon e^{(m+1)}) \wedge \ldots \wedge (e^m + i_\varepsilon e^{2m}) \qquad (2.48)$$

for a constant $a \in \mathbb{R}^*$.

In dimension 6, there is a characterisation of $\mathrm{SU}^\varepsilon(p, q)$-structures in terms of stable forms. Given a six-dimensional real vector space V, we call a pair (ω, ρ) of a stable $\omega \in \Lambda^2 V^*$ and a stable $\rho \in \Lambda^3 V^*$ compatible if it holds

$$\omega \wedge \rho = 0. \tag{2.49}$$

Let us recall, that the stabiliser in $GL(V)$ of a compatible pair is

$$\mathrm{Stab}_{GL(V)}(\omega, \rho) = SU^\varepsilon(p, q), \quad p + q = 3,$$

where $\varepsilon \in \{\pm 1\}$ is the sign of $\lambda(\rho)$, that is $J_\rho^2 = \varepsilon id_V$. This is seen as follows. For the two-form ω, stability is equivalent to non-degeneracy and we choose the orientation on V such that ω^3 is positive. By the previous section, we can associate an ε-complex structure J_ρ to the stable three-form ρ. For instance in an adequate basis, it is easy to verify that $\omega \wedge \rho = 0$ is equivalent to the skew-symmetry of J_ρ with respect to ω. Equivalently, the pseudo-Euclidean metric

$$g = \varepsilon \, \omega(\cdot, J_\rho \cdot), \tag{2.50}$$

induced by ω and ρ is ε-Hermitian with respect to J_ρ. Since $\Psi_\rho = \rho + i_\varepsilon J_\rho^* \rho$ is a $(3, 0)$-form and the stabiliser of ω and ρ also stabilises the tensors induced by them, the claim follows.

We conclude that an $SU^\varepsilon(p, q)$-structure, $p + q = 3$, on a six-manifold is characterised by a pair $(\omega, \psi^+) \in \Omega^2 M \times \Omega^3 M$ of everywhere stable and compatible forms such that the induced $(3, 0)$-form $\Psi = \psi^+ + i_\varepsilon J_{\psi^+}^* \psi^+ = \psi^+ + i_\varepsilon \psi^-$ has constant non-zero length with respect to the induced metric (2.50). In an ε-unitary frame which is adapted to Ψ in the sense of (2.48), the formula

$$\psi^- \wedge \psi^+ = \|\psi^+\|^2 \frac{1}{6} \omega^3 \tag{2.51}$$

is easily verified. Thus, given a compatible pair (ω, ψ^+) of stable forms, it can be checked that the induced $(3, 0)$-form Ψ has constant non-zero length without explicitly computing the induced metric.

We remark, that in the almost Hermitian case, the literature often requires Ψ to be normalised such that $\|\psi^+\|^2 = 4$, for instance in [32]. In the more general almost ε-Hermitian case, we have in an adapted local ε-unitary frame (2.48) with $\sigma_1 = \sigma_2$

$$\psi^+ = a(e^{123} + \varepsilon(e^{156} + e^{426} + e^{453})) \quad \text{and} \quad \|\psi^+\|^2 = 4a^2 \sigma_3 \tag{2.52}$$

for a real constant a. Therefore we have to consider two different normalisations $\|\psi^+\| = \pm 4$ or we multiply the metric by -1 if necessary such that $\|\psi^+\|^2$ is always positive.

Finally, we remark that $SU(3)$-structures are classified in [32] and it is shown that the intrinsic torsion is completely determined by the exterior derivatives $d\omega$, $d\psi^+$ and $d\psi^-$.

2.5   Pseudo-Riemannian Submersions

Let us consider the setting of a pseudo-Riemannian submersion $\pi : (M, g) \to (N, h)$. The tangent bundle TM of M splits orthogonally into the direct sum

$$TM = \mathcal{H} \oplus \mathcal{V}, \tag{2.53}$$

where $\mathcal{V}$ is the vertical and $\mathcal{H}$ the horizontal distribution. Denote by $\iota_{\mathcal{H}}, \iota_{\mathcal{V}}$ the canonical inclusions and by $\pi_{\mathcal{H}}, \pi_{\mathcal{V}}$ the canonical projections. We recall the definition [18, 100] of the fundamental tensorial invariants A and T of the submersion π

$$T_\zeta = \pi_{\mathcal{H}} \circ \nabla_{\pi_{\mathcal{V}}(\zeta)} \circ \pi_{\mathcal{V}} + \pi_{\mathcal{V}} \circ \nabla_{\pi_{\mathcal{V}}(\zeta)} \circ \pi_{\mathcal{H}},$$

$$A_\zeta = \pi_{\mathcal{H}} \circ \nabla_{\pi_{\mathcal{H}}(\zeta)} \circ \pi_{\mathcal{V}} + \pi_{\mathcal{V}} \circ \nabla_{\pi_{\mathcal{H}}(\zeta)} \circ \pi_{\mathcal{H}},$$

where ζ is a vector field on M.

The components of the Levi-Civita connection ∇ are given in the next proposition (compare [100], [18] 9.24 and 9.25).

Proposition 2.5.1 *Let* $\pi : (M, g) \to (N, h)$ *be a pseudo-Riemannian submersion, denote by* ∇ *the Levi-Civita connection of* g *and define* $\nabla^{\mathcal{V}} := \pi_{\mathcal{V}} \circ \nabla \circ \iota_{\mathcal{V}}$. *For vector fields* X, Y *in* $\mathcal{H}$ *and* U, V *in* $\mathcal{V}$ *we have the following identities*

$$\nabla_U V = \nabla_U^{\mathcal{V}} V + T_U V, \tag{2.54}$$

$$\nabla_U X = T_U X + \pi_{\mathcal{H}}(\nabla_U X), \tag{2.55}$$

$$\nabla_X U = \pi_{\mathcal{V}}(\nabla_X U) + A_X U, \tag{2.56}$$

$$\nabla_X Y = A_X Y + \pi_{\mathcal{H}}(\nabla_X Y), \tag{2.57}$$

$$\pi_{\mathcal{V}}[X, Y] = 2A_X Y, \tag{2.58}$$

$$g(A_X Y, U) = -g(A_X U, Y), \text{ or more generally } A \text{ is alternating.} \tag{2.59}$$

The canonical variation of the metric g for $t \in \mathbb{R} - \{0\}$ is given by

$$g_t := \begin{cases} g(X, Y), \text{ for } X, Y \in \mathcal{H}, \\ tg(V, W), \text{ for } V, W \in \mathcal{V}, \\ g(V, X) = 0, \text{ for } V \in \mathcal{V}, X \in \mathcal{H}. \end{cases}$$

Lemma 2.5.2 *Denote by* X, Y *vector fields in* $\mathcal{H}$ *and by* U, V *vector fields in* $\mathcal{V}$.

(1) Let A^t *and* T^t *be the tensorial invariants for* g_t *and* A *and* T *those for* $g = g_1$. *Then it holds*

$$A_X^t Y = A_X Y, \quad A_X^t U = t A_X U \text{ and} \tag{2.60}$$

$$T_U^t V = t T_U V, \quad T_U^t X = T_U X; \tag{2.61}$$

(2) $\nabla_U^{tV} V = \nabla_U^V V$;

(3)

$$\pi_{\mathcal{H}}(\nabla_X^t Y) = \pi_{\mathcal{H}}(\nabla_X Y) \text{ and } \pi_{\mathcal{V}}(\nabla_X^t V) = \pi_{\mathcal{V}}(\nabla_X V); \qquad (2.62)$$

(4)

$$\pi_{\mathcal{H}}(\nabla^t{}_V X - \nabla_V X) = (t-1) A_X V; \qquad (2.63)$$

$$\nabla_U^t V = \nabla_U V + (t-1) T_U V. \qquad (2.64)$$

Proof The first part can be found in Lemma 9.69 of [18]. On the right hand-side of the Koszul formulas one only needs the metric g_t on $\mathcal{V}$ to determine $\nabla^{t\mathcal{V}}$. This shows $\nabla^{t\mathcal{V}} = \nabla^{\mathcal{V}}$. An analogous argument using the Koszul formulas shows $\pi_{\mathcal{H}}(\nabla^t{}_X Y) = \pi_{\mathcal{H}}(\nabla_X Y)$ and $\pi_{\mathcal{V}}(\nabla_X^t V) = \pi_{\mathcal{V}}(\nabla_X V)$. The first part of the point (4) follows from the identities (2.58) and (2.59) and the Koszul formulas. The last equation follows from (1) and (2): $\nabla_U^t V = \nabla_U^{tV} V + T_U^t V = \nabla_U^V V + t T_U V = \nabla_U V + (t-1) T_U V$. □

2.6 Para-Sasaki Manifolds

In Sect. 3.7 of Chap. 3 we give a construction of (non-flat) Ricci-flat (strict) nearly para-Kähler manifolds using a cone construction over para-Sasaki manifolds, which we are therefore going to discuss now. Let (M, h) be a semi-Riemannian manifold. A cone is a semi-Riemannian manifold

$$(\widehat{M} = \mathbb{R}^+ \times M, g_c := cdr^2 + r^2 h) \text{ with } c \in \mathbb{R}^*,$$

where $r > 0$ is the coordinate of $\mathbb{R}^+$. One calls g_c space-like for $c > 0$ and time-like for $c < 0$. Since we are considering semi-Riemannian metrics, we may suppose $c = \pm 1$ by rescaling g_c. For lifts of vector fields X, Y, Z, U on M to $\widehat{M}$ the curvature tensor[4] $\widehat{R}$ of $\widehat{M}$ is

$$\widehat{R}(X, Y, Z, U) = r^2 \left(R(X, Y, Z, U) - \frac{1}{c} [g(Y, Z)g(X, U) - g(X, Z)g(Y, U)] \right).$$

Moreover, it is

$$\partial_r \lrcorner \widehat{R} = 0.$$

Hence $\widehat{R}$ is flat if and only if R has constant curvature $\frac{1}{c}$. As in the complex setting there are different characterisations of the para-complex analogue of Sasaki

---

[4]See for example [100, pp. 209–211] for the curvature of warped products.

manifolds and different names occur in the literature like para-Sasaki manifolds [8] or time-like Sasaki manifolds [82].

Definition 2.6.1 (See Definition 6.3 of [82] or Proposition 8.2 of [8])

A para-Sasaki manifold is a triple (M, g, T), where (M, g) is a semi-Riemannian manifold of signature[5] $(m, m + 1)$ and T is a time-like unit Killing vector field such that the endomorphism $\Phi := DT \in \Gamma(\text{End}(TM))$ satisfies

$$\Phi^2 = Id + g(\cdot, T)T, \tag{2.65}$$

$$(D_X\Phi)Y = -g(X, Y)T + g(Y, T)X, \quad \forall X, Y \in \Gamma(TM). \tag{2.66}$$

A para-Sasaki Einstein manifold (M, g, T) is a para-Sasaki manifold such that g is an Einstein metric.

Remarks 2.6.2

(i) For a semi-Riemannian manifold (M, g) there is a one-to-one correspondence (Theorem 8.1 of [8]) between para-Sasaki structures (M, g, T) and para-Kähler structures on the (space-like) cone $(\widehat{M}, \widehat{g} = g_1, \widehat{P})$ with

$$\widehat{P}(\partial_r) = \frac{T}{r}, \ \widehat{P}(T) = r\partial_r \text{ and } \widehat{P}(X) = \widehat{D}_X T = D_X^{\widehat{g}} T, \text{ for } X \in T^\perp.$$

(ii) The Einstein equation for a para-Sasaki manifold (M, g, T) of dimension $2m+1$ implies as in the Riemannian case that the scalar curvature equals $2m(2m + 1)$ or the Einstein constant equals $2m$ (see Proposition 6.5 of [82]).

In particular, the corresponding para-Kähler cone $(\widehat{M}, \widehat{g}, \widehat{P})$ is Ricci-flat if and only if (M, g, T) is para-Sasaki Einstein. Therefore M is a para-Sasaki Einstein manifold if and only if the (restricted) holonomy group of its cone $\widehat{M}$ is contained in $SU(P_0, g_0)$ (see Proposition 6.6 of [82]), which by Remark 3.1.12 (ii) of Chap. 3 stabilises an m-form that for $m = 3$ has isotropic support (in the sense of Sect. 2.3). In consequence $\widehat{M}$ can be endowed with a parallel m-form and in dimension 6, one case of our interest, $\widehat{M}$ admits a parallel three-form of isotropic support.

2.6.1 The T-Dual Space

In order to explain the examples given below we recall the concept of the T-dual space introduced in [81, 82].

Let $\mathfrak{g}$ be a compact Lie algebra and $\mathfrak{h} \subset \mathfrak{g}$ be a sub-algebra. One fixes an $ad(\mathfrak{h})$-invariant complement $\mathfrak{m}$ of $\mathfrak{h}$ in $\mathfrak{g}$ and an involutive automorphism with

$$T : \mathfrak{g} \to \mathfrak{g}, T \neq Id.$$

---

[5]In our convention $m + 1$ is the number of negative directions.

Let

$$\mathfrak{g}_\pm := \{X \in \mathfrak{g} \mid TX = \pm X\}$$

be the ± 1-eigenspaces of T. This data gives rise to another (non-compact) real form

$$\mathfrak{g}' := \mathfrak{g}_+ \oplus i\mathfrak{g}_-$$

of the complexified Lie algebra $\mathfrak{g}^{\mathbb{C}}$. One additionally fixes an $ad(\mathfrak{h})$-invariant scalar product $\langle \cdot, \cdot \rangle$ on $\mathfrak{m}$. Further one assumes that it holds

$$T(\mathfrak{h}) = \mathfrak{h} \text{ and } T(\mathfrak{m}) = \mathfrak{m}$$

and that $T : \mathfrak{m} \to \mathfrak{m}$ is an isometry.

Then

$$\mathfrak{h}' = \mathfrak{h} \cap \mathfrak{g}_+ \oplus i(\mathfrak{h} \cap \mathfrak{g}_-) \subset \mathfrak{g}'$$

is a sub-algebra and setting $\mathfrak{m}' := \mathfrak{m} \cap \mathfrak{g}_+ \oplus i(\mathfrak{m} \cap \mathfrak{g}_-)$ one has

$$\mathfrak{g}' = \mathfrak{m}' \oplus \mathfrak{h}' \text{ and } [\mathfrak{h}', \mathfrak{m}'] \subset \mathfrak{m}'.$$

The restriction $\langle \cdot, \cdot \rangle'$ of the sesquilinear extension on $\mathfrak{m}^{\mathbb{C}}$ of $\langle \cdot, \cdot \rangle$ to $\mathfrak{m}' \subset \mathfrak{m}^{\mathbb{C}}$ yields an $ad(\mathfrak{h}')$-invariant pseudo-Euclidean scalar product $\langle \cdot, \cdot \rangle'$.

Definition 2.6.3 The quadruple $(\mathfrak{g}', \mathfrak{h}', \mathfrak{m}', \langle \cdot, \cdot \rangle')$ constructed from $(\mathfrak{g}, \mathfrak{h}, \mathfrak{m}, \langle \cdot, \cdot \rangle)$ in the above way is called *T-dual* to $(\mathfrak{g}, \mathfrak{h}, \mathfrak{m}, \langle \cdot, \cdot \rangle)$.

Definition 2.6.4 Let G/H be a Riemannian homogeneous space, where G is a compact Lie group. Denote by $(\mathfrak{g}, \mathfrak{h}, \mathfrak{m}, \langle \cdot, \cdot \rangle)$ the associated data, then the semi-Riemannian homogeneous space G'/H' with associated data $(\mathfrak{g}', \mathfrak{h}', \mathfrak{m}', \langle \cdot, \cdot \rangle')$ is called T-dual if $(\mathfrak{g}, \mathfrak{h}, \mathfrak{m}, \langle \cdot, \cdot \rangle)$ and $(\mathfrak{g}', \mathfrak{h}', \mathfrak{m}', \langle \cdot, \cdot \rangle')$ are T-dual for some involutive automorphism T of $\mathfrak{g}$.

Examples 2.6.5 As mentioned in the Introduction we obtain a correspondence (Theorems 3.7.3 and 3.7.5 of Chap. 3) between conical Ricci-flat (strict) nearly para-Kähler manifolds and conical Ricci-flat para-Kähler manifolds. Regular conical Ricci-flat para-Kähler manifolds can be characterised as local cones over para-Sasaki Einstein manifolds (Proposition 3.7.8 of Chap. 3). In particular, the cone over a para-Sasaki Einstein manifold yields a conical Ricci-flat para-Kähler manifold and via our construction a conical Ricci-flat (strict) nearly para-Kähler manifold. For this reason we include the subsequent classes of examples.

(i) This class of examples is given in Example 6.2 of [82]: Denote by (B, h, P) a $2m$-dimensional para-Kähler Einstein manifold with scalar curvature

$4m(m + 1)$ and consider the $\mathbb{R}^*$-bundle[6] $M = P_{U(P_0, g_0)} \times_\rho \mathbb{R}^*$, where ρ is the following representation of the para-unitary group

$$\rho : U(P_0, g_0) \to \mathbb{R}^*, \quad \begin{pmatrix} A & 0 \\ 0 & (A^t)^{-1} \end{pmatrix} \mapsto det(A)$$

with the connection φ induced by the Levi-Civita connection. On M we define a metric by

$$g(X, Y) := (\pi^* h)(X, Y) - \frac{1}{(m + 1)^2} \varphi(X) \varphi(Y), \text{ for } X, Y \in \Gamma(TM),$$

where the Lie algebra of $\mathbb{R}^*$ is identified with $\mathbb{R}$ and where $\pi : M \to B$ is the projection. Moreover, let $\eta := \frac{1}{m+1} \varphi$, T the vector field defined by $\eta = g(T, \cdot)$, then (M, g, T) is a para-Sasaki Einstein manifold.

(ii) Consider a compact homogeneous space $M = G/H$ with left-invariant Riemannian metric g associated to $(\mathfrak{g}, \mathfrak{h}, \mathfrak{m}, \langle \cdot, \cdot \rangle)$. Let (ξ, η, φ) be a Sasakian structure on (M, g), which is G-invariant. Moreover, let $(G'/H', g')$ be T-dual to $(G/H, g)$, where H' is connected and denote by ξ_0 the $Ad\, H$-invariant vector in $\mathfrak{m}$ corresponding to the G-invariant vector field ξ. Provided that $\xi_0 \in \mathfrak{m}_-$, then $i\xi_0$ defines a G'-invariant vector field ξ' on G'/H'. Setting $\Phi := -D\xi'$ and $\eta' := g(\xi', \cdot)$ one obtains a para-Sasaki structure on $(G'/H', g')$, cf. Example 6.3 of [82]. If g is an Einstein metric, then by Corollary 4.2 of [81] the T-dual metric g' is an Einstein metric, too.

Let us recall that a compact homogeneous Sasaki-Einstein manifold M of dimension 5 is a circle bundle over either $\mathbb{C}P^2$ or $\mathbb{C}P^1 \times \mathbb{C}P^1$ which means that M is either S^5 or $S^2 \times S^3$, see Corollary 11.1.14 of [20]. The Einstein metric on $S^2 \times S^3$ was constructed in [79] (see also Section 10.2 of [56]) by considering it as Stiefel manifold $V_{4,2} = SO(4)/SO(2)$ and the related T-duals are discussed in Section 8.1 of [82]. The T-dual space of importance here is $SO^+(2, 2)/SO^+(1, 1)$ which fibers over

$$SO^+(2, 1)/SO^+(1, 1) \times SO^+(2, 1)/SO^+(1, 1)$$

and is para-Sasaki Einstein of non constant sectional curvature, since from this manifold we get an (first) example for a non-flat Ricci-flat nearly para-Kähler manifold. This structure has also recently been considered in Section 3 of [29] as the natural generalisation of the well-known Sasaki-Einstein metric on the tangent sphere bundle $T_1 S^3 \cong S^3 \times S^2$.

(iii) Examples of para-Sasaki Einstein manifolds also arise in projective special para-Kähler geometry, see for instance Theorem 4.4.13 of Chap. 4 or [39].

---

[6] Note, that the frame bundle of a para-Kähler manifold can be reduced to the bundle of para-unitary frames $P_{U(P_0, g_0)}$.

Chapter 3
Nearly Pseudo-Kähler and Nearly Para-Kähler Manifolds

3.1 Nearly Pseudo-Kähler and Nearly Para-Kähler Manifolds

3.1.1 General Properties

In this subsection we collect some information on almost ε-Hermitian manifolds with a special emphasis on the nearly ε-Kähler case.

Definition 3.1.1 An almost ε-Hermitian manifold $(M^{2m}, g, J^\varepsilon, \omega)$ is called nearly ε-Kähler manifold, provided that its Levi-Civita connection ∇ satisfies the nearly ε-Kähler condition

$$(\nabla_X J^\varepsilon) X = 0, \quad \forall X \in TM.$$

A nearly ε-Kähler manifold is called strict if $\nabla_X J^\varepsilon \neq 0$ for all non-trivial vector fields X.

A tensor field $B \in \Gamma((TM^*)^{\otimes 2} \otimes TM)$ on a pseudo-Riemannian manifold (M, g) is called (totally) skew-symmetric if the tensor $g(B(X, Y), Z)$ is a three-form. The following characterisation of a nearly ε-Kähler manifold is well-known in the Riemannian context and we refer to Proposition 3.2 of [110] for the complete proof in the pseudo-Riemannian setting.

Proposition 3.1.2 An almost ε-Hermitian manifold $(M^{2m}, g, J^\varepsilon, \omega)$ satisfies the nearly ε-Kähler condition if and only if $d\omega$ is of real type $(3, 0) + (0, 3)$ and the Nijenhuis tensor is totally skew-symmetric.

Remark 3.1.3 The notion of nearly ε-Kähler manifold corresponds to the generalised Gray-Hervella class $\mathcal{W}_1$ in [86]. However, in the para-Hermitian case, there

© Springer International Publishing AG 2017

L. Schäfer, *Nearly Pseudo-Kähler Manifolds and Related Special Holonomies*,
Lecture Notes in Mathematics 2201, DOI 10.1007/978-3-319-65807-0_3

are two subclasses, see [59]. Indeed, we already observed that

$$\mathcal{A} = -\nabla\omega \in [\![\Omega^{3,0}]\!] \overset{(2.36)}{=} \Gamma(\Lambda^3\mathcal{V}^* \oplus \Lambda^3\mathcal{H}^*)$$

for a nearly para-Kähler manifold.

Definition 3.1.4 A connection $\bar{\nabla}$ on an ε-Hermitian manifold $(M^{2m}, g, J^\varepsilon, \omega)$ is called ε-**Hermitian** provided, that it satisfies $\bar{\nabla}g = 0$ and $\bar{\nabla}J^\varepsilon = 0$.

The Riemannian case of the next result is due to [57], the para-complex case is shown in [78]. In fact, the sketched proof in [57] holds literally for the almost pseudo-Hermitian case with indefinite signature as well. A direct and simultaneous proof of all cases can be found in Proposition 3.4 of [110].

Theorem 3.1.5 *An ε-Hermitian manifold $(M^{2m}, g, J^\varepsilon, \omega)$ admits an ε-Hermitian connection with totally skew-symmetric torsion if and only if the Nijenhuis tensor is totally skew-symmetric. If this is the case, the connection $\bar{\nabla}$ and its torsion T are uniquely defined by*

$$g(\bar{\nabla}_X Y, Z) = g(\nabla_X Y, Z) + \frac{1}{2}g(T(X, Y), Z),$$

$$g(T(X, Y), Z) = \varepsilon g(N(X, Y), Z) - d\omega(J^\varepsilon X, J^\varepsilon Y, J^\varepsilon Z),$$

and we call $\bar{\nabla}$ the characteristic ε-Hermitian connection (with skew-symmetric torsion).

This connection can be seen as a natural generalisation of the Chern- or Bismut-connection. Another name for the characteristic connection is canonical connection.

Remark 3.1.6 An almost Hermitian manifold is said to be of type $\mathcal{G}_1$ if it admits a Hermitian connection with skew-symmetric torsion. In terms of the Gray-Hervella list [68], this means, that it is of type $\mathcal{W}_1 \oplus \mathcal{W}_3 \oplus \mathcal{W}_4$, i.e. the missing part is the almost Kähler component $\mathcal{W}_2$.

More generally, the proposition justifies to say that an almost ε-Hermitian manifold is of type $\mathcal{G}_1$ if it admits an ε-Hermitian connection with skew-symmetric torsion.

In particular, the proposition applies to nearly ε-Kähler manifolds $(M, g, J^\varepsilon, \omega)$. In fact, comparing the identities (2.35) and (2.41), we see that the real three-form $\mathcal{A}$ is of type $(3, 0) + (0, 3)$. Since $d\omega$ is the alternation of $\nabla\omega$, we have

$$d\omega = 3\nabla\omega = -3\mathcal{A} \in [\![\Omega^{3,0}]\!], \qquad (3.1)$$

where $\mathcal{A}$ is defined in (2.39). Furthermore, if we apply the nearly ε-Kähler condition to the expression (2.37), the Nijenhuis tensor of a nearly ε-Kähler structure simplifies to

$$N(X, Y) = 4J^\varepsilon(\nabla_X J^\varepsilon)Y. \qquad (3.2)$$

We conclude that the Nijenhuis tensor is skew-symmetric since

$$g(N(X, Y), Z) = -4\mathcal{A}(X, Y, J^\varepsilon Z) \overset{(2.41)}{=} -4\varepsilon J^{\varepsilon *} \mathcal{A}(X, Y, Z). \tag{3.3}$$

Explicitly the connection $\bar{\nabla}$ is then given by

$$\bar{\nabla}_X Y = \nabla_X Y + \frac{1}{2}\varepsilon J^\varepsilon (\nabla_X J^\varepsilon) Y, \text{ for } X, Y \in \Gamma(TM). \tag{3.4}$$

In this case, the skew-symmetric torsion T of the characteristic ε-Hermitian connection simplifies to

$$T(X, Y) = \varepsilon J^\varepsilon (\nabla_X J^\varepsilon) Y = \frac{1}{4}\varepsilon N(X, Y)$$

due to the identities (3.1)–(3.3).

For a proof of the next result we may refer to Lemma 2.4 of [16] for nearly Kähler manifolds, Theorem 5.3 of [78] for nearly para-Kähler manifolds and Proposition 3.2 of [108] for the remaining case. As the attentive reader observes, the proof relies on the curvature identity (3.20), even though we list it already in this section as one of the very useful properties of the characteristic connection.

Proposition 3.1.7 *The characteristic ε-Hermitian connection $\bar{\nabla}$ of a nearly ε-Kähler manifold $(M^{2m}, J^\varepsilon, g, \omega)$ satisfies*

$$\bar{\nabla}(\nabla J^\varepsilon) = 0 \quad and \quad \bar{\nabla}(T) = 0.$$

A direct consequence is the following Corollary.

Corollary 3.1.8 *On a nearly ε-Kähler manifold $(M^{2m}, J^\varepsilon, g, \omega)$ the tensors ∇J^ε and $N = 4\varepsilon T$ have constant length.*

Remark 3.1.9 In dimension 6, the fact that ∇J^ε has constant length is usually expressed by the equivalent assertion that a nearly ε-Kähler six-manifold is of constant type, i. e. there is a constant $\alpha \in \mathbb{R}$ such that

$$g((\nabla_X J^\varepsilon)Y, (\nabla_X J^\varepsilon)Y) = \alpha \{ g(X, X)g(Y, Y) - g(X, Y)^2 + \varepsilon g(J^\varepsilon X, Y)^2 \}. \tag{3.5}$$

In fact, the constant is $\alpha = \frac{1}{4}\|\nabla J^\varepsilon\|^2$. Furthermore, it is well-known in the Riemannian case that strict nearly Kähler six-manifolds are Einstein manifolds with Einstein constant 5α [67]. The same is true in the para-Hermitian case [78] and in the pseudo-Hermitian case [108] or Theorem 3.2.8 of this chapter. The sign of the type constant depends on the signature $(2p, 2q)$ of g by $sign(p-q)$, see for example [82]. In particular, in the Riemannian case it follows $\alpha > 0$ and as a consequence a strict nearly Kähler manifold cannot be Ricci-flat.

The case $\|\nabla J^\varepsilon\|^2 = 0$ for a strict nearly ε-Kähler six-manifold can only occur in the para-complex world. We give different characterisations of such structures which provide an obvious break in the analogy of nearly para-Kähler and nearly pseudo-Kähler manifolds. To emphasise that we are only considering the nearly para-Kähler case we write τ for J^ε with $\varepsilon = 1$.

Proposition 3.1.10 *For a six-dimensional strict nearly para-Kähler manifold (M^6, g, τ, ω) the following properties are equivalent:*

(i) $\|\nabla\tau\|^2 = \|\mathcal{A}\|^2 = 0$.
(ii) *The three-form* $\mathcal{A} = -\nabla\omega \in [\![\Omega^{3,0}]\!]$ *is either in* $\Gamma(\Lambda^3 \mathcal{V}^*)$ *or in* $\Gamma(\Lambda^3 \mathcal{H}^*)$.
(iii) *The three-form* $\mathcal{A} = -\nabla\omega \in [\![\Omega^{3,0}]\!]$ *is not stable.*
(iv) *The metric g is Ricci-flat.*

In consequence for a Ricci-flat nearly para-Kähler manifold the 3-forms $D\omega(\cdot, \cdot, \cdot)$ and $N(\cdot, \cdot, \cdot)$ are not stable in the sense of Hitchin [75, 76], cf. Sect. 2.1 of Chap. 2 for details on stable forms and hence the powerful methods of stable forms are not available. The following observation is used later in this text to construct examples of non-flat Ricci-flat nearly para-Kähler six-manifolds.

Corollary 3.1.11

(a) *On a Ricci-flat nearly para-Kähler six-manifold (M^6, τ, g) the 3-forms $\nabla\omega$ and N have isotropic support.*
(b) *Let (M, τ, g) be a nearly para-Kähler manifold such that the Nijenhuis tensor N has isotropic support, then one has $N_X \circ N_Y = 0$.*

Proof The identity (3.5) combined with $\alpha = 0$ yields $g((\nabla_X\tau)Y, (\nabla_X\tau)Y) = 0$ and further

$$g(\tau(\nabla_X\tau)Y, \tau(\nabla_X\tau)Y) = 0. \tag{3.6}$$

This shows that the two 3-forms $g((\nabla_X\tau)Y, Z)$ and $g(\tau(\nabla_X\tau)Y, Z)$ have isotropic support. Finally we obtain after polarisation of (3.6), that one has

$$g(N_X N_Y Z, W) = -16g(\tau(\nabla_Y\tau)Z, \tau(\nabla_X\tau)W) = 0$$

for all $X, Y, Z, W \in \Gamma(TM)$. This yields the last statement.   □

Remark 3.1.12 Let us consider $\mathbb{R}^{2m}$ with its standard para-Hermitian structure (P_0, g_0) and isotropic basis $(e_1, \ldots, e_m, f_1, \ldots, f_m)$ with dual isotropic basis $(e^1, \ldots, e^m, f^1, \ldots, f^m)$, compare Eq. (2.33). Then the m-forms $e^1 \wedge \ldots \wedge e^m$ and $f^1 \wedge \ldots \wedge f^m$ are invariant under $SU(P_0, g_0)$, which follows from Eq. (2.34) and have isotropic support in the above sense for $m = 3$.

As the (restricted) holonomy of a Ricci-flat para-Kähler six-manifold (M^6, P, g) lies in $SU(P_0, g_0)$, it follows that a Ricci-flat para-Kähler six-manifold admits a family of (non-vanishing) parallel 3-forms with isotropic support.

Lemma 3.1.13 *Let (M, τ, g) be a nearly para-Kähler manifold such that the Nijenhuis tensor $N(X, Y, Z)$ has isotropic support, then it holds*

$$(\nabla_X N)(Y, Z) = 0, \tag{3.7}$$

$$(\nabla_X N)(Y, Z, W) = 0. \tag{3.8}$$

In particular, these identities are satisfied for a Ricci-flat nearly para-Kähler six-manifold.

Proof We directly compute Eq. (3.7) using Theorem 3.1.5 and Corollary 3.1.11 (b)

$$(\nabla_X N)(Y, Z) = \nabla_X (N(Y, Z)) - N(\nabla_X Y, Z) - N(Y, \nabla_X Z)$$

$$= \left(\bar{\nabla}_X - \frac{1}{8} N_X \right) (N(Y, Z)) - N \left(\left(\bar{\nabla}_X - \frac{1}{8} N_X \right) Y, Z \right)$$

$$- N \left(Y, \left(\bar{\nabla}_X - \frac{1}{8} N_X \right) Z \right) = (\bar{\nabla}_X N)(Y, Z) = 0.$$

Combining Eq. (3.7) with $\nabla g = 0$ and $N(X, Y, Z) = g(N(X, Y), Z)$ we obtain Eq. (3.8). The last statement follows from Corollary 3.1.11 (a).   □

Flat strict nearly para-Kähler manifolds (M, g, J, ω) are classified in work with V. Cortés, see Sect. 3.6 of this chapter. It turns out that these always satisfy $\|\nabla J^\varepsilon\|^2 = 0$. In [59], almost para-Hermitian structures on tangent bundles TN of real three-dimensional manifolds N^3 are discussed. It is shown that the existence of nearly para-Kähler manifolds satisfying the second condition of Proposition 3.1.10 is equivalent to the existence of a certain connection on N^3 without constructing an example. However, to our best knowledge, there was no reference for an example of a Ricci-flat non-flat strict nearly para-Kähler structure until the author's paper [109] discussed in Sect. 3.7 of this chapter.

3.1.2 Characterisations by Exterior Differential Systems in Dimension 6

The following lemma explicitly relates the Nijenhuis tensor to the exterior differential. For $\varepsilon = -1$, it gives a characterisation of Bryant's notion of a quasi-integrable $U(p, q)$-structure, $p + q = 3$, in dimension 6 [24].

Let $(M^6, g, J^\varepsilon, \omega)$ be a six-dimensional almost ε-Hermitian manifold. If $\{e_1, \ldots, e_6 = Je_3\}$ is a local ε-unitary frame, we define a local frame $\{E^1, E^2, E^3\}$ of $(TM^{1,0})^*$ by

$$E^i := (e^i + \mathrm{i}_\varepsilon \varepsilon J^\varepsilon e^i) = (e^i + \mathrm{i}_\varepsilon e^{i+m})$$

for $i = 1, 2, 3$ and call it a local ε-unitary frame of $(1, 0)$-forms. The dual vector fields of the $(1, 0)$-forms are

$$E_i = e_i^{1,0} = \frac{1}{2}(e_i + i_\varepsilon \varepsilon J^\varepsilon e_i) = \frac{1}{2}(e_i + i_\varepsilon \varepsilon e_{i+m}),$$

such that the $\mathbb{C}_\varepsilon$-bilinearly extended metric in this kind of frame satisfies

$$g(E_i, \bar{E}_j) = \frac{1}{2}\sigma_i \delta_{ij} \qquad \text{and} \qquad g(E_i, E_j) = 0.$$

Lemma 3.1.14 *The Nijenhuis tensor of an almost ε-Hermitian six-manifold $(M^6, g, J^\varepsilon, \omega)$ is totally skew-symmetric if and only if for every local ε-unitary frame of $(1, 0)$-forms, there exists a local $\mathbb{C}_\varepsilon$-valued function λ such that*

$$(dE^{s(1)})^{0,2} = \lambda \, \sigma_{s(1)} \, E^{\overline{s(2)}\,\overline{s(3)}} \tag{3.9}$$

for all even permutations s of $\{1, 2, 3\}$.

Proof First of all, the identities

$$N(\bar{V}, \bar{W}) = -4\varepsilon[\bar{V}, \bar{W}]^{1,0} \quad \text{and} \quad N(V, \bar{W}) = 0$$

for any vector fields $V = V^{1,0}$, $W = W^{1,0}$ in $TM^{1,0}$ follow immediately from the definition of N. Using the first identity, we compute in an arbitrary local ε-unitary frame

$$dE^i(\bar{E}_j, \bar{E}_k) = -E^i([\bar{E}_j, \bar{E}_k]) = -2\sigma_i \, g([\bar{E}_j, \bar{E}_k], \bar{E}_i)$$

$$= -2\sigma_i \, g([\bar{E}_j, \bar{E}_k]^{1,0}, \bar{E}_i) = \frac{1}{2}\varepsilon \, \sigma_i \, g(N(\bar{E}_j, \bar{E}_k), \bar{E}_i)$$

for all possible indices $1 \leq i, j, k \leq 3$. If the Nijenhuis tensor is totally skew-symmetric, Eq. (3.9) follows by setting

$$\lambda = \frac{1}{2}\varepsilon \, g(N(\bar{E}_1, \bar{E}_2), \bar{E}_3). \tag{3.10}$$

Conversely, the assumption (3.9) for every local ε-unitary frame implies that the Nijenhuis tensor is everywhere a three-form when considering the same computation and $N(V, \bar{W}) = 0$.   □

From the last Lemma we get the following Corollary.

Corollary 3.1.15 *For an almost ε-Hermitian six-manifold $(M^6, g, J^\varepsilon, \omega)$ with totally skew-symmetric Nijenhuis tensor, there exists a function $f \in C^\infty(M)$ such that one has*

$$g_p(X, Y) = f(p) \, \mathrm{tr}(N_X \circ N_Y), \, p \in M, \, X, Y \in T_p M.$$

In particular, if this function f does not vanish, i.e. if the almost complex structure is quasi-integrable, the almost ε-complex structure fixes the conformal class of g.

If there is an $SU^\varepsilon(p,q)$-reduction (cf. Sect. 2.4 of Chap. 2) with closed real part, this characterisation can be reformulated globally in the following sense.

Proposition 3.1.16 *Let (ω, ψ^+) be an $SU^\varepsilon(p,q)$-structure on a six-manifold M such that ψ^+ is closed. Then the Nijenhuis tensor is totally skew-symmetric if and only if*

$$d\psi^- = v\,\omega \wedge \omega \tag{3.11}$$

for a global real function v.

Proof It suffices to proof this locally. Let $\{E^i\}$ be an ε-unitary frame of $(1,0)$-forms with $\sigma_1 = \sigma_2$ which is adapted to the $SU^\varepsilon(p,q)$-reduction such that $\Psi = \psi^+ + i_\varepsilon \psi^- = a E^{123}$ for a real constant a as in (2.48). The fundamental two-form is

$$\omega = -\frac{1}{2} i_\varepsilon \sum_{k=1}^{m} \sigma_k E^{k\bar{k}}$$

in such a frame. Furthermore, as ψ^+ is closed, we have $d\Psi = i_\varepsilon d\psi^- = -d\bar{\Psi}$, which implies that $d\Psi \in \Lambda^{2,2}$. Considering this, we compute the real 4-form

$$d\psi^- = \varepsilon i_\varepsilon\, d\Psi = \varepsilon i_\varepsilon a \left((dE^1)^{0,2} \wedge E^{23} + (dE^2)^{0,2} \wedge E^{31} + (dE^3)^{0,2} \wedge E^{12} \right)$$

and compare this expression with

$$\omega \wedge \omega = \frac{1}{2} \varepsilon (\sigma_2\sigma_3\, E^{2\bar{2}3\bar{3}} + \sigma_1\sigma_3\, E^{1\bar{1}3\bar{3}} + \sigma_1\sigma_2\, E^{1\bar{1}2\bar{2}})$$

$$= -\frac{1}{2}\varepsilon\sigma_3(\sigma_1\, E^{\bar{2}3\bar{3}2\bar{3}} + \sigma_2\, E^{\bar{3}131} + \sigma_3\, E^{\bar{1}\bar{2}12}).$$

Hence, by Lemma 3.1.14, the Nijenhuis tensor is totally skew-symmetric if and only if $d\psi^- = v\,\omega \wedge \omega$ holds true for a real function v. More precisely, the two functions v and λ are related by the formula

$$v = -2\sigma_3 i_\varepsilon a\lambda. \tag{3.12}$$

$\square$

An $SU^\varepsilon(p,q)$-structure (ω, ψ) is called *half-flat* if

$$d\psi = 0, \quad d\omega^2 = 0,$$

and *nearly half-flat* if

$$d\psi = v\,\omega \wedge \omega$$

for a real constant v. These notions are defined for the Riemannian signature in [32] respectively [53] and extended to all signatures in our paper [46] presented in Chap. 4 of this text.

Corollary 3.1.17 *Let (ω, ψ^+) be a half-flat $SU^\varepsilon(p, q)$-structure on a six-manifold M. Then, the Nijenhuis tensor is totally skew-symmetric if and only if (ω, ψ^-) is nearly half-flat.*

Proof If (ω, ψ^-) is nearly half-flat, Eq. (3.11) is satisfied by definition and the Nijenhuis tensor is skew-symmetric by the previous proposition. In particular one has $d\omega^2 = 0$. Conversely, if the Nijenhuis tensor is skew, we know that (3.11) holds true for a real function v, since we have $d\psi^+ = 0$. Differentiating this equation and using $d\omega^2 = 0$, we obtain $dv \wedge \omega^2 = 0$. The assertion follows as wedging by ω^2 is injective on one-forms.   □

Remark 3.1.18 An interesting property of $SU^\varepsilon(p, q)$-structures which are both half-flat and nearly half-flat in the sense of the corollary is the fact that, given that the manifold and the $SU^\varepsilon(p, q)$-structure are analytic, the structure can be evolved to both a parallel G_2-structure and a nearly parallel G_2-structure via the Hitchin flow. For details, we refer to [76] and [114] for the compact Riemannian case and Chap. 4 of this text or our paper [46] for the non-compact case and indefinite signatures.

In [33], six-dimensional nilmanifolds N admitting an invariant half-flat $SU(3)$-structure (ω, ψ^+) such that (ω, ψ^-) is nearly half-flat are classified. As six nilmanifolds admit such a structure, we conclude that these structures are not as scarce as nearly Kähler manifolds. It is also shown in the same article, that these structures induce invariant G_2-structures with torsion on $N \times S^1$.

We give another example of a (normalised) left-invariant $SU(3)$-structure on $S^3 \times S^3$ which satisfies $d\psi^+ = 0$, $d\psi^- = \omega \wedge \omega$ such that $d\omega$ neither vanishes nor is of type $(3,0) + (0,3)$. We choose a global frame of left-invariant vector fields $\{e_1, e_2, e_3, f_1, f_2, f_3\}$ on $S^3 \times S^3$ such that

$$de^1 = e^{23}, \quad de^2 = e^{31}, \quad de^3 = e^{12}; \quad df^1 = f^{23}, \quad df^2 = f^{31}, \quad df^3 = f^{12},$$

and set with $x = 2 + \sqrt{3}$

$$\omega = e^1 f^1 + e^2 f^2 + e^3 f^3,$$

$$\psi^+ = -\frac{1}{2}x^2 e^{123} + 2xe^{12}f^3 - 2xe^{13}f^2 - 2xe^1 f^{23} + 2xe^{23}f^1$$
$$+ 2xe^2 f^{13} - 2xe^3 f^{12} + (4x - 8)f^{123},$$

$$\psi^- = \frac{1}{2}xe^{123} - 2e^1 f^{23} + 2e^2 f^{13} - 2e^3 f^{12} + 4f^{123},$$

$$g = x(e^1)^2 + x(e^2)^2 + x(e^3)^2 + 4(f^1)^2 + 4(f^2)^2 + 4(f^3)^2$$
$$- 2xe^1 \cdot f^1 - 2xe^2 \cdot f^2 - 2xe^3 \cdot f^3.$$

Finally, we come to the characterisation of six-dimensional nearly ε-Kähler manifolds by an exterior differential system generalising the classical result of [103] which holds for $\varepsilon = -1$ and Riemannian metrics.

Theorem 3.1.19 *Let $(M, g, J^\varepsilon, \omega)$ be an almost ε-Hermitian six-manifold. Then M is a strict nearly ε-Kähler manifold with $\|\nabla J^\varepsilon\|^2 \neq 0$ if and only if there is a reduction $\Psi = \psi^+ + i_\varepsilon \psi^-$ to $\mathrm{SU}^\varepsilon(p, q)$ which satisfies*

$$d\omega = 3\,\psi^+, \tag{3.13}$$

$$d\psi^- = 2\,\alpha\,\omega \wedge \omega, \tag{3.14}$$

where $\alpha = \frac{1}{4}\|\nabla J^\varepsilon\|^2$ is constant and non-zero.

Remark 3.1.20 Due to our sign convention $\omega = g(., J^\varepsilon .)$, the constant α is positive in the Riemannian case and the second equation differs from that of other authors. Furthermore, we will sometimes use the term nearly ε-Kähler manifold of non-zero type if $\|\nabla J^\varepsilon\|^2 \neq 0$.

Proof By Proposition 3.1.2, the manifold M is nearly ε-Kähler if and only if $d\omega$ is of type $(3, 0) + (0, 3)$ and the Nijenhuis tensor is totally skew-symmetric.

Therefore, when $(g, J^\varepsilon, \omega)$ is a strict nearly ε-Kähler structure such that $\|\mathcal{A}\|^2 = \|\nabla J^\varepsilon\|^2$ is constant (by Corollary 3.1.8) and not zero (by assumption), we can define the reduction $\Psi = \psi^+ + i_\varepsilon \psi^-$ by $\psi^+ = \frac{1}{3} d\omega = -\mathcal{A}$ and $\psi^- = J^{\varepsilon*}\psi^+$ such that the first equation is satisfied. Since ω is of type $(1, 1)$ and therefore $d(\omega \wedge \omega) = 2d\omega \wedge \omega = 0$, this reduction is half-flat. Thus, Corollary 3.1.17 and the skew-symmetry of N imply that there is a constant $\nu \in \mathbb{R}$ such that $d\psi^- = \nu\,\omega \wedge \omega$.

According to (2.48), we can choose an ε-unitary local frame with $\sigma_1 = \sigma_2$, such that

$$\Psi = -\mathcal{A} - i_\varepsilon J^{\varepsilon*}\mathcal{A} = aE^{123},$$

where a is constant and satisfies $4\alpha = \|\nabla J^\varepsilon\|^2 = \|\psi^+\|^2 = 4a^2\sigma_3$ by (2.52). Now, the functions defined in Lemma 3.1.14 and Proposition 3.1.16 evaluate as

$$\lambda \stackrel{(3.10)}{=} \frac{1}{2}\varepsilon g(N(\bar{E}_1, \bar{E}_2), \bar{E}_3) \stackrel{(3.3)}{=} -2J^*\mathcal{A}(\bar{E}_1, \bar{E}_2, \bar{E}_3) = -\varepsilon\,i_\varepsilon a,$$

$$\nu \stackrel{(3.12)}{=} -2\sigma_3 i_\varepsilon a\lambda = 2\sigma_3 a^2 = 2\alpha.$$

Conversely, if a given $\mathrm{SU}^\varepsilon(p, q)$-structure satisfies the exterior system, the real three-form ψ^+ is obviously closed and the Nijenhuis tensor is totally skew-symmetric by Corollary 3.1.17. Considering that $d\omega = 3\nabla\omega$ is of type $(3, 0)+(0, 3)$ by the first equation, the structure is nearly ε-Kähler. Since $\mathcal{A} = -\psi^+$ is stable, the structure is strict nearly ε-Kähler and $\|\nabla J^\varepsilon\| = \|\mathcal{A}\| \neq 0$ by Proposition 3.1.10. Now, the computation of the constants in the adapted ε-unitary frame shows that in fact $\|\nabla J^\varepsilon\| = 4\alpha$. $\qquad\square$

3.1.3 Curvature Identities for Nearly ε-Kähler Manifolds

Most of these identities are here only used for the almost complex case. If we are only considering the complex case we write J and in case, that we consider the para-complex case we write τ for the ε-complex structure J^ε. The starting point of a series of curvature identities are

$$R(W,X,Y,Z) - R(W,X,JY,JZ) = g((\nabla_W J)X, (\nabla_Y J)Z), \qquad (3.15)$$

$$R(W,X,W,Z) + R(W,JX,W,JZ) \qquad\qquad\qquad\qquad (3.16)$$
$$- R(W,JW,X,JZ) = 2g((\nabla_W J)X, (\nabla_W J)Z),$$

$$R(W,X,Y,Z) = R(JW,JX,JY,JZ), \qquad (3.17)$$

which were already proven for pseudo-Riemannian metrics by Gray [67]. In the para-complex case the analogue of the first identity, i.e. the relation

$$R(W,X,Y,Z) + R(W,X,\tau Y,\tau Z) = g((\nabla_W \tau)X, (\nabla_Y \tau)Z), \qquad (3.18)$$

is shown in Proposition 5.2 of [78].

Let $\{e_i\}_{i=1}^{2n}$ be a local orthonormal frame field, then the Ricci- and the Ricci*-tensor are given by

$$g(\operatorname{Ric} X, Y) = \sum_{i=1}^{2n} \epsilon_i R(X, e_i, Y, e_i), \quad g(\operatorname{Ric}^* X, Y) = \frac{1}{2} \sum_{i=1}^{2n} \epsilon_i R(X, JY, e_i, Je_i)$$

with $\epsilon_i = g(e_i, e_i) = g(Je_i, Je_i)$ and $X, Y \in TM$. The frame $\{e_i\}_{i=1}^{2n}$ is called **adapted** if it holds $Je_i = e_{i+n}$ for $i = 1, \ldots, n$. Then it follows using an adapted frame from Eqs. (3.16) and (3.17) that

$$g(rX, Y) := g((\operatorname{Ric} - \operatorname{Ric}^*)X, Y) = \sum_{i=1}^{2n} \epsilon_i g((\nabla_X J)e_i, (\nabla_Y J)e_i). \qquad (3.19)$$

Using the right hand-side we see

$$[J, r] = 0.$$

For the second derivative of the complex structure one has the identity

$$2g(\nabla_{W,X}^2(J)Y, Z) = -\sigma_{X,Y,Z}\, g((\nabla_W J)X, (\nabla_Y J)JZ), \qquad (3.20)$$

which was proven in [67] for Riemannian metrics and holds true in the pseudo-Riemannian setting, cf. [82, Proposition 7.1]. This identity implies

$$\sum_{i=1}^{2n} \epsilon_i \nabla^2_{e_i,e_i}(J)Y = -r(JY). \tag{3.21}$$

From Proposition 3.1.7 and the relation (3.4) of the connections ∇ and $\bar{\nabla}$ one obtains the following identities for the curvature tensor $\bar{R}$ of $\bar{\nabla}$ and the curvature tensor R of the Levi-Civita connection ∇

$$\bar{R}(W,X,Y,Z) = R(W,X,Y,Z) - \frac{1}{2}g((\nabla_W J)X, (\nabla_Y J)Z)$$

$$+ \frac{1}{4}\left[g((\nabla_W J)Y, (\nabla_X J)Z) - g((\nabla_W J)Z, (\nabla_X J)Y)\right] \tag{3.22}$$

$$= \frac{1}{4}[3R(W,X,Y,Z) + R(W,X,JY,JZ)$$

$$+ \sigma_{XYZ}R(W,X,JY,JZ)],$$

$$\bar{R}(W,JW,Y,JZ) = \frac{1}{4}[5R(W,JW,Y,JZ)$$

$$- R(W,Y,W,Z) - R(W,JY,W,JZ)]. \tag{3.23}$$

With the help of Eq. (3.22) it follows

$$\bar{R}(W,X,Y,Z) = \bar{R}(Y,Z,W,X) = -\bar{R}(X,W,Y,Z) = -\bar{R}(W,X,Z,Y). \tag{3.24}$$

Using $\bar{\nabla}J = 0$ and $\bar{\nabla}g = 0$ we obtain

$$\bar{R}(W,X,Y,Z) = \bar{R}(W,X,JY,JZ) \tag{3.25}$$

$$= \bar{R}(JW,JX,Y,Z) = \bar{R}(JW,JX,JY,JZ).$$

The general form of the first Bianchi identity (cf. Chapter III of [87]) for a connection with torsion yields in the case of parallel torsion:

$$\sigma_{XYZ} \bar{R}(W,X,Y,Z) = -\sigma_{XYZ} g((\nabla_W J)X, (\nabla_Y J)Z). \tag{3.26}$$

In a similar way we get from the second Bianchi identity (cf. Chapter III of [87]) for a connection with parallel torsion or from the second Bianchi identity for ∇

$$- \sigma_{VWX} \bar{\nabla}_V(\bar{R})(W,X,Y,Z) = \sigma_{VWX} \bar{R}((\nabla_V J)JW, X, Y, Z). \tag{3.27}$$

From deriving Eq. (3.22) and the second Bianchi identity of ∇ one gets after a direct computation

$$\underset{VWX}{\sigma} \nabla_V(\bar{R})(W, X, Y, Z) = \frac{1}{2}g((\nabla_Y J)Z, \underset{VWX}{\sigma} (\nabla_X J)(\nabla_V J)JW), \qquad (3.28)$$

which implies

$$\underset{VWX}{\sigma} \nabla_V(\bar{R})(W, X, Y, JY) = 0. \qquad (3.29)$$

Proposition 3.1.21 (Proposition 2.3 of [108]) *The tensor r on a nearly pseudo-Kähler manifold (M, J, g) is parallel with respect to the characteristic connection $\bar{\nabla}$.*

Theorem 3.1.22 (Theorem 2.4 of [108]) *Let (M, J, g) be a nearly pseudo-Kähler manifold and let W, X be vector fields on M then it holds*

$$\sum_{i,j=1}^{2n} \epsilon_i \epsilon_j g(re_i, e_j) \left[R(W, e_i, X, e_j) - 5R(W, e_i, JX, Je_j) \right] = 0. \qquad (3.30)$$

Let us remark, that the Riemannian case is done in [67] and the para-Kähler case in [78].

3.2 Structure Results

As we have seen above, for a nearly pseudo-Kähler manifold $\nabla \omega$ is a differential form of type $(3, 0) + (0, 3)$. In consequence real two- or four-dimensional nearly pseudo-Kähler manifolds are automatically pseudo-Kähler. Six dimensional nearly pseudo-Kähler manifolds are either pseudo-Kähler manifolds or strict nearly pseudo-Kähler manifolds. Therefore we start this section in real dimension 8.[1]

3.2.1 Kähler Factors and the Structure in Dimension 8

The aim of this subsection is to split off the pseudo-Kähler factor of a nearly pseudo-Kähler manifold. This will be done by means of the kernel of ∇J and allows to reduce the (real) dimension from 8 to 6.

For $p \in M$ we set

$$\mathcal{K}_p = \ker(X \in T_p M \mapsto \nabla_X J).$$

---

[1]The reference for the section is the author's paper [108].

Theorem 3.2.1 *Let (M, J, g) be a nearly pseudo-Kähler manifold. Suppose, that the distribution $\mathcal{K}$ has constant dimension and admits an orthogonal complement,*

(i) *then M is locally a pseudo-Riemannian product $M = K \times M_1$ of a pseudo-Kähler manifold K and a strict nearly pseudo-Kähler manifold M_1.*

(ii) *if M is complete and simply connected then it is a pseudo-Riemannian product $M = K \times M_1$ of a pseudo-Kähler manifold K and a strict nearly pseudo-Kähler manifold M_1.*

Proof The distribution $\mathcal{K}$ is parallel for the characteristic connection $\bar{\nabla}$, since ∇J is $\bar{\nabla}$-parallel. By the formula (3.4) and the nearly Kähler condition it follows $\bar{\nabla}_X K = \nabla_X K$ for sections K in $\mathcal{K}$ and X in TM. This implies that $\mathcal{K}$ is parallel for the Levi-Civita connection and in consequence its orthogonal complement $(\mathcal{K})^\perp$ is Levi-Civita parallel. The proof of (i) finishes by the local version of the theorem of de Rham and the proof of (ii) by the global version. $\qquad\square$

Remark 3.2.2 There exist nearly pseudo-Kähler manifolds (M, J, g) without pseudo-Kähler de Rham factor, such that $\mathcal{K}_\eta \neq \{0\}$ admits no orthogonal complement. In fact, we construct Levi-Civita flat nearly pseudo-Kähler manifolds in our paper [41], which is subject of Sect. 3.6 of this chapter, such that the three-form $\eta_p(X, Y, Z) = g_p(J(\nabla_X J)Y, Z)$, for $p \in M$, has a support $\Sigma_\eta \subset T_p M$ which is a maximally isotropic subspace (Here we identified $T_p M$ and $T_p^* M$ via the metric g.). Obviously, $J(\nabla_X J)Y$ and $J(\nabla_U J)V$ are elements of the support of η for arbitrary $X, Y, U, V \in T_p M$. It then follows $0 = g(J(\nabla_X J)Y, J(\nabla_U J)V) = g(J(\nabla_{J(\nabla_X J)Y} J)U, V)$ for all $V \in T_p M$. Hence it is $\Sigma_\eta \subset \mathcal{K}_\eta$. Moreover for general reasons we have shown before $\Sigma_\eta = \mathcal{K}_\eta^\perp$ which shows $\mathcal{K}_\eta \cap \mathcal{K}_\eta^\perp \neq \{0\}$ for the above examples. From these examples we learn, that the Theorem 3.2.1 does not hold true, if there is no orthogonal complement.

Definition 3.2.3 A nearly pseudo-Kähler manifold (M, J, g) is called nice if the three-form $g((\nabla . J^\varepsilon)\cdot, \cdot)$ has non-zero length in each point $p \in M$.

Theorem 3.2.4 *Let (M^8, J, g) be a complete simply connected eight-dimensional nice nearly pseudo-Kähler manifold. Then $M = M_1 \times M_2$ where M_1 is a two-dimensional Kähler manifold and M_2 is a six-dimensional strict nearly pseudo-Kähler manifold.*

Proof Since (M, J, g) is a nice nearly pseudo-Kähler manifold we can use Lemma 2.3.6 of Chap. 2 to obtain an orthogonal splitting in the two-dimensional distribution $\mathcal{K}$ and its orthogonal complement, which coincides with Σ_η. Therefore we are in the situation of Theorem 3.2.1 (ii). $\qquad\square$

3.2.2 Einstein Condition Versus Reducible Holonomy

In this part we study reducible $\bar{\nabla}$-holonomy and discuss the consequences in small dimensions.

Theorem 3.2.5 *Let (M, J, g) be a nearly pseudo-Kähler manifold.*

(i) *Suppose that r has more than one eigenvalue, then the characteristic Hermitian connection has reduced holonomy.*

(ii) *If the tensor field r has exactly one eigenvalue then M is a pseudo-Riemannian Einstein manifold.*

Proof

(i) Let μ_i for $i = 1, \ldots, l$ be the eigenvalues of r. Then the decomposition in the according eigenbundles $\mathrm{Eig}(\mu_i)$ is $\bar{\nabla}$-parallel and hence its holonomy is reducible.

(ii) From the identity of Theorem 3.1.22 and $r = \mu \mathrm{Id}_{TM}$ we obtain

$$0 = \sum_{i=1}^{2n} \epsilon_i \left(R(W, e_i, X, e_i) - 5R(W, e_i, JX, Je_i) \right) = g((Ric - 5Ric^*)W, X),$$

where we used the Bianchi identity and an adapted frame to obtain the last equality. This shows comparing with $r = Ric - Ric^*$ that it holds $Ric = \frac{5}{4}\mu$.

□

Let us recall, that in the pseudo-Riemannian setting the decomposition into the eigenbundles is **not** automatically ensured to be an orthogonal direct decomposition. Therefore we introduce the following notion.

Definition 3.2.6 A nearly pseudo-Kähler manifold (M, J, g) is called decomposable if the above decomposition into the eigenbundles of the tensor r is orthogonal.

Lemma 3.2.7 *Let (M, J, g) be a decomposable nearly pseudo-Kähler manifold and denote by μ_i for $i = 1, \ldots, l$ the eigenvalues of r and by $E_i = \mathrm{Eig}(\mu_i), i = 1, \ldots, l$, the corresponding eigenbundles.*

(i) *For $X \in E_i$ and $Y \in E_j$ with $i \neq j$ one has $Ric(X, Y) = 0$.*

(ii) *For $X, Y \in E_i$ it is*

$$Ric(X, Y) = \frac{\mu_i}{4} g(X, Y) + \frac{1}{\mu_i} \sum_{s=1}^{l} \mu_s \, g(r^s X, Y),$$

where the tensors $r^s : TM \to TM$, $1 \leq s \leq l$, are defined as

$$g(r^s X, Y) := -\mathrm{tr}_{E_s} \left((\nabla_X J) \circ (\nabla_Y J) \right).$$

Proof Let us first prove (i). We consider a basis of TM which gives a pseudo-orthonormal basis for the $E_i, i = 1, \ldots l$. The Ricci curvature decomposes w.r.t. the eigenbundles as

$$Ric(X, Y) = \sum_{s=1}^{l} \sum_{e_k \in E_s} \epsilon_k R(X, e_k, Y, e_k).$$

Using $\bar{R}(X, e_k, Y, e_k) = 0$ for $s \neq j$ one gets by Eq. (3.22)

$$R(X, e_k, Y, e_k) = \frac{1}{4}g((\nabla_X J)e_k, (\nabla_Y J)e_k), \text{ for } e_k \in E_s.$$

Further for $s = j$ one has $s \neq i$ and again it is

$$R(X, e_k, Y, e_k) = R(Y, e_k, X, e_k) = \frac{1}{4}g((\nabla_X J)e_k, (\nabla_Y J)e_k), \text{ for } e_k \in E_s.$$

In summary one obtains

$$Ric(X, Y) = \sum_{s=1}^{l} \sum_{e_k \in E_s} \epsilon_k R(X, e_k, Y, e_k) = \frac{1}{4}\sum_{k} \epsilon_k g((\nabla_X J)e_k, (\nabla_Y J)e_k) = \frac{1}{4}g(rX, Y) = 0.$$

This shows (i). Next we show part (ii). From the identity of Theorem 3.1.22 we conclude

$$0 = \sum_{s=1}^{l} \sum_{e_k \in E_s} \mu_s \epsilon_k \left(R(W, e_k, X, e_k) - 5R(W, e_k, JX, Je_k) \right).$$

As in part (i) we get for $s \neq i$ with help of Eq. (3.22)

$$R(X, e_k, JY, Je_k) = -3R(Y, e_k, X, e_k) = -\frac{3}{4}g((\nabla_X J)e_k, (\nabla_Y J)e_k), \text{ for } e_k \in E_s.$$

It follows, that

$$4\sum_{s \neq i} \mu_s g(r^s X, Y) + \mu_i \left(\sum_{e_k \in E_i} \epsilon_k \left(R(W, e_k, X, e_k) - 5R(W, e_k, JX, Je_k) \right) \right) = 0$$

and another time using Eq. (3.22)

$$\mu_i g((Ric - 5Ric^*)X, Y) + 4\sum_{s \neq i} (\mu_s - \mu_i)g(r^s X, Y) = 0,$$

which follows by

$$g((Ric - 5Ric^*)X, Y) = \sum_{s=1}^{l} \sum_{e_k \in E_s} \epsilon_k \left(R(X, e_k, Y, e_k) - 5R(X, e_k, JY, Je_k) \right).$$

The identity (ii) follows now from $Ric - Ric^* = r$ and $\sum_{s=1}^{l} r^s = r$. In fact, it is

$$g((Ric - 5Ric^*)X, Y) = -4g(Ric\,X, Y) + 5g(rX, Y) = \frac{4}{\mu_i} \sum_{s \neq i} (\mu_i - \mu_s) g(r^s X, Y)$$

and in consequence one obtains

$$4g(Ric\,X, Y) = 5g(rX, Y) - \frac{4}{\mu_i} \sum_{s \neq i} (\mu_i - \mu_s) g(r^s X, Y) = g(rX, Y) + \frac{4}{\mu_i} \sum_{s=1}^{l} \mu_s g(r^s X, Y),$$

which finishes the proof.   □

Theorem 3.2.8 *A strict nearly pseudo-Kähler six-manifold (M^6, J, g) of constant type α is a pseudo-Riemannian Einstein manifold with Einstein constant 5α.*

Proof In an adapted basis we obtain from the symmetries of ∇J

$$g(rX, X) = 2 \sum_{i=1}^{3} \epsilon_i\, g((\nabla_X J)e_i, (\nabla_X J)e_i) = -2 \sum_{i=1}^{3} \epsilon_i\, g((\nabla_X J)^2 e_i, e_i).$$

This is exactly minus the trace of the operator $(\nabla_X J)^2$ which has a simple form in a cyclic frame. It follows after polarising $g(rX, Y) = 4\alpha g(X, Y)$. From Theorem 3.2.5 we compute the Einstein constant 5α where α is the type constant of the strict nearly pseudo-Kähler manifold M^6.   □

Proposition 3.2.9 *Let (M^{10}, J, g) be a nice nearly pseudo-Kähler ten-manifold.*

(i) *Then the tensor r in a frame of the first type in Lemma 2.3.8 of Chap. 2 is given by*

$$re_1 = 4(\alpha^2 + \beta^2)e_1,$$
$$re_2 = 4\alpha^2 e_2, \quad re_3 = 4\alpha^2 e_3,$$
$$re_4 = 4\beta^2 e_4, \quad re_5 = 4\beta^2 e_5,$$
$$r(Je_i) = Jr(e_i), \quad i = 1, \ldots, 5,$$

where α, β are constants.

(ii) *For a frame of the second type in Lemma 2.3.8 of Chap. 2 the tensor r is given by*

$$r \begin{bmatrix} e_1 \\ e_2 \\ e_3 \end{bmatrix} = 4 \begin{bmatrix} \alpha^2 + \beta^2 \epsilon_4 \epsilon_5 & 0 & \beta^2 \epsilon_4 \epsilon_5 \\ 0 & \alpha^2 & 0 \\ \beta^2 \epsilon_4 \epsilon_5 & 0 & \alpha^2 + \beta^2 \epsilon_4 \epsilon_5 \end{bmatrix} \begin{bmatrix} e_1 \\ e_2 \\ e_3 \end{bmatrix}$$

$$re_4 = 0,$$

$$re_5 = 4\beta^2(2\epsilon_1\epsilon_4 - 1)e_5,$$

$$r(Je_i) = Jr(e_i), \quad i = 1, \ldots, 5.$$

The eigenvalues are $\{0; 4\alpha^2; 4\beta^2(2\epsilon_1\epsilon_4 - 1); 4(\alpha^2 + 2\beta^2\epsilon_4\epsilon_5)\}$, *where the eigenbundles are given as*

$$\mathrm{Ker}(r) = \mathrm{span}\{e_4, Je_4\},$$

$$\mathrm{Eig}(r, 4\alpha^2) = \mathrm{span}\{-e_1 + e_3, e_2, -Je_1 + Je_3, Je_2\},$$

$$\mathrm{Eig}(r, 4\beta^2(2\epsilon_1\epsilon_4 - 1)) = \mathrm{span}\{e_5, Je_5\},$$

$$\mathrm{Eig}(r, 4(\alpha^2 + 2\beta^2\epsilon_4\epsilon_5))) = \mathrm{span}\{e_1 + e_3, Je_1 + Je_3\},$$

where α, β *are constants. For* $\beta^2 \neq 0$ *the second case is not decomposable.*
(iii) *Suppose* $\beta = 0$ *in the cases (i) and (ii). Then it follows*

$$\mathrm{Eig}(r, 4\alpha^2) = \mathrm{span}\{e_1, e_2, e_3, Je_1, Je_2, Je_3\},$$

$$\mathrm{Ker}(r) = \mathrm{span}\{e_4, e_5, Je_4, Je_5\}.$$

Proof In an adapted frame we obtain from the symmetries of ∇J

$$g(rX, Y) = 2\sum_{i=1}^{5} \epsilon_i g((\nabla_X J)e_i, (\nabla_Y J)e_i) = -2\sum_{i=1}^{5} \epsilon_i g((\nabla_Y J)(\nabla_X J)e_i, e_i).$$

This is exactly minus the trace of the operator $(\nabla_Y J)(\nabla_X J)$. Using the form of Lemma 2.3.8 of Chap. 2 one can calculate r by hand or using computer algebra systems to obtain the claimed results $\qquad\square$

Theorem 3.2.10 *Let* (M^{10}, J, g) *be a complete simply connected nice decomposable nearly pseudo-Kähler manifold of dimension 10. Then* M^{10} *is of one of the following types*

(i) *the tensor r has a kernel and* $M^{10} = K \times M^6$ *is a product of a four-dimensional pseudo-Kähler manifold K and a strict nearly pseudo-Kähler six-manifold* M^6.
(ii) *the tensor r has trivial kernel and r has eigenvalues* $4(\alpha^2 + \beta^2)$ *with multiplicity* 2, $4\alpha^2, 4\beta^2$ *with multiplicity 4 for some* $\alpha, \beta \neq 0$.

A nice nearly pseudo-Kähler manifold (M^{10}, J, g) *is decomposable if the dimension of the kernel of r is not equal to two.*

Proof Since we suppose, that (M^{10}, J, g) is a nice and decomposable nearly pseudo-Kähler manifold, Proposition 3.2.9 implies that one has the two different cases:

(i) the distribution $\mathcal{K}$, which is the tangent space of the Kähler factor has dimension 4 and admits an orthogonal complement of dimension 6. This is part (iii) of Proposition 3.2.9. Part (i) of the Theorem now follows from Theorem 3.2.1.

(ii) the tensor r has trivial kernel and we are in the situation of Proposition 3.2.9
part (i) with $\alpha, \beta \neq 0$ and part (ii) follows.

$\square$

Remark 3.2.11 Nearly pseudo-Kähler manifolds falling in the second case of the
last theorem are related to twistor spaces in Sect. 3.4.5 of this chapter.

3.3 Twistor Spaces over Quaternionic and Para-Quaternionic Kähler Manifolds

In this section[2] we consider pseudo-Riemannian submersions $\pi : (M, g) \rightarrow$
(N, h) endowed with a complex structure J on M which is compatible with the
decomposition (2.53).

Lemma 3.3.1 (Lemma 5.1 of [108]) *Let* $\pi : (M, g) \rightarrow (N, h)$ *be a pseudo-
Riemannian submersion endowed with a complex structure* J *on* M *which is
compatible with the decomposition (2.53). Then* (M, g, J) *is a pseudo-Kähler
manifold if and only if the following equations*[3] *are satisfied*

$$\pi_{\mathcal{H}}((\nabla_X J)Y) = \pi_{\mathcal{H}}((\nabla_V J)X) = 0, \tag{3.31}$$

$$(\nabla_U^{\mathcal{V}} J)V = \pi_{\mathcal{V}}((\nabla_X J)V) = 0, \tag{3.32}$$

$$A_X(JY) - JA_X Y = 0, \quad A_X(JV) - JA_X V = 0, \tag{3.33}$$

$$T_V(JX) - JT_V X = 0, \quad T_U(JV) - JT_U V = 0, \tag{3.34}$$

where X, Y *are vector fields in* $\mathcal{H}$ *and* U, V *are vector fields in* $\mathcal{V}$.

Further we define a second complex structure by

$$\hat{J} := \begin{cases} J \text{ on } \mathcal{H}, \\ -J \text{ on } \mathcal{V}. \end{cases}$$

We observe that $\hat{\hat{J}} = J$. This construction was made in [98] for the Riemannian
setting and imitates the construction on twistor spaces, which was first done in [50].

Proposition 3.3.2 *Suppose, that the foliation induced by the pseudo-Riemannian
submersion* π *is totally geodesic and that* (M, J, g) *is a pseudo-Kähler manifold and
J is compatible with the decomposition (2.53), then the manifold* $(M, \hat{g} = g_{\frac{1}{2}}, \hat{J})$ *is a
nearly pseudo-Kähler manifold. The distributions* $\mathcal{H}$ *and* $\mathcal{V}$ *are parallel with respect*

---

[2]The reference still is the author's paper [108].

[3]Please note, the small difference between the torsion $T(X, Y)$ of $\bar{\nabla}$ and the second fundamental
form $T_U V$, which is not dangerous, as later we are considering totally geodesic fibrations.

to the characteristic Hermitian connection $\bar{\nabla}$ of $(M, \hat{g}, \hat{J})$. In other words the nearly pseudo-Kähler manifold $(M, \hat{g}, \hat{J})$ has reducible $\bar{\nabla}$-holonomy.

Proof Let U, V be vector fields in $\mathcal{V}$ and X, Y be vector fields in $\mathcal{H}$: In the following $\hat{\nabla}$ is the Levi-Civita connection of $\hat{g}$. Since the fibres are totally geodesic, i.e. $T \equiv 0$, we obtain from Eq. (2.54), that $\hat{\nabla}_U V = \hat{\nabla}_U^{\mathcal{V}} V + \hat{T}_U V = \nabla_U^{\mathcal{V}} V + T_U V = \nabla_U V$, which yields $(\hat{\nabla}_U \hat{J}) V = -(\nabla_U J) V = 0$.

In the sequel we denote the O'Neill tensors of the pseudo-Riemannian foliations induced by $\mathcal{V}$ on (M, g) and on $(M, \hat{g})$ by A and $\hat{A}$, respectively. From Lemma 2.5.2 of Chap. 2 it follows $A_X Y = \hat{A}_X Y$ and consequently the same Lemma yields $\nabla_X Y = \hat{\nabla}_X Y$.

Since (M, g) is Kähler, Lemma 3.3.1 implies $A \circ J = J \circ A$ and we compute

$$
\begin{aligned}
(\hat{\nabla}_X \hat{J}) Y \quad &= \quad \hat{\nabla}_X (\hat{J} Y) - \hat{J} \hat{\nabla}_X Y \qquad\qquad\qquad\qquad (3.35) \\
&= \quad \pi_{\mathcal{H}}[\hat{\nabla}_X (JY)] + \pi_{\mathcal{V}}[\hat{\nabla}_X (JY)] - \hat{J}(\pi_{\mathcal{H}}(\hat{\nabla}_X Y) + \pi_{\mathcal{V}}(\hat{\nabla}_X Y)) \\
&= \quad \pi_{\mathcal{H}}[\hat{\nabla}_X (JY) - J \hat{\nabla}_X Y] + \pi_{\mathcal{V}}[\hat{\nabla}_X (JY) + J \hat{\nabla}_X Y] \\
&= \quad \pi_{\mathcal{H}}((\hat{\nabla}_X J) Y) + \hat{A}_X (JY) + J \hat{A}_X Y \\
&\overset{(2.60),(2.62),(3.33)}{=} \pi_{\mathcal{H}}((\nabla_X J) Y) + 2 A_X (JY) \overset{(3.31)}{=} 2 A_X (JY) = 2 J A_X Y.
\end{aligned}
$$

With the identity $A_X V = 2 \hat{A}_X V$ of Lemma 2.5.2 of Chap. 2 we get

$$
\begin{aligned}
(\hat{\nabla}_X \hat{J}) V \quad &= \quad \hat{\nabla}_X (\hat{J} V) - \hat{J} \hat{\nabla}_X V \qquad\qquad\qquad\qquad (3.36) \\
&= \quad -\pi_{\mathcal{V}}(\hat{\nabla}_X (JV)) - \pi_{\mathcal{H}}(\hat{\nabla}_X (JV)) + J \pi_{\mathcal{V}}(\hat{\nabla}_X V) - J \pi_{\mathcal{H}}(\hat{\nabla}_X V) \\
&= \quad -\pi_{\mathcal{V}}((\hat{\nabla}_X J) V) - \hat{A}_X JV - J \hat{A}_X V \\
&\overset{(2.60),(2.62),(3.33)}{=} -\pi_{\mathcal{V}}((\nabla_X J) V) - J A_X V = -A_X JV.
\end{aligned}
$$

The vanishing of the second fundamental form T, Eq. (2.61) and a second time $A_X V = 2 \hat{A}_X V$ show

$$
\begin{aligned}
(\hat{\nabla}_V \hat{J}) X \quad &= \quad \pi_{\mathcal{V}}(\hat{\nabla}_V (JX)) + \pi_{\mathcal{V}}(J \hat{\nabla}_V X) + \pi_{\mathcal{H}}(\hat{\nabla}_V (JX) - J \hat{\nabla}_V X) \qquad (3.37) \\
&\overset{(2.63)}{=} \hat{T}_V (JX) + J(\hat{T}_V X) + \pi_{\mathcal{H}}((\nabla_V J) X) + \frac{1}{2}(J A_X V - A_{JX} V) = J A_X V,
\end{aligned}
$$

where we used $A_{JX} V = -J A_X V$ which follows, since A_X is alternating (compare Eq. (2.59)) and commutes with J. The next Lemma finishes the proof. $\qquad\square$

Lemma 3.3.3 (Lemma 5.3 of [108])

1) *Suppose, that $(M, \hat{J}, \hat{g})$ is a nearly pseudo-Kähler manifold and $\hat{J}$ is compatible with the decomposition (2.53), then the following statements are equivalent:*

   (i) *the splitting (2.53) is $\bar{\nabla}$-parallel,*

(ii) *the fundamental tensors $\hat{A}$ and $\hat{T}$ satisfy:*

$$\hat{T}_V X = 0, \quad \hat{J}\hat{T}_V W = -\hat{T}_V \hat{J} W \Leftrightarrow \check{J}\hat{T}_V W = \hat{T}_V \check{J} W \, for \, \check{J} = \hat{J}, \quad (3.38)$$

$$\hat{A}_X V = \frac{1}{2}\hat{J}(\hat{\nabla}_X \hat{J})V, \quad \hat{A}_X Y = \frac{1}{2}\pi_V \left(\hat{J}(\hat{\nabla}_X \hat{J})Y \right). \quad (3.39)$$

2) *If it holds $(\hat{\nabla}_V \hat{J})W = 0$ then $\bar{\nabla}_V W \in \mathcal{V}$ for $V, W \in \mathcal{V}$ is equivalent to $T_V W = 0$. Moreover it is $(\hat{\nabla}_V^\mathcal{V} \hat{J})W = 0$.*

We apply Proposition 3.3.2 to twistor spaces and obtain.

Corollary 3.3.4 *The twistor space $\mathcal{Z}$ of a quaternionic Kähler manifold of dimension $4k$ with negative scalar curvature admits a canonical nearly pseudo-Kähler structure of reducible holonomy contained in $U(1) \times U(2k)$.*

Proof We remark, that in negative scalar curvature the twistor space of a quaternionic Kähler manifold is the total space of a pseudo-Riemannian submersion with totally geodesic fibres. It admits a compatible pseudo-Kähler structure of signature $(2, 4k)$, cf. Besse [18, 14.86 b)]. The assumption of positive scalar curvature is often made to obtain a positive definite metric on $\mathcal{Z}$. Here we focus on pseudo-Riemannian metrics and consequently on negative scalar curvature. $\qquad\Box$

Proposition 3.3.5 *The twistor spaces $\mathcal{Z}$ of non-compact duals of Wolf spaces and of Alekseevskian spaces admit a nearly pseudo-Kähler structure.*

Proof Non-compact duals of Wolf spaces are known [117] to be quaternionic Kähler manifolds of negative scalar curvature. The same holds for Alekseevskian spaces [3, 38]. $\qquad\Box$

Studying the lists given in [3, 38, 117] we find examples of six-dimensional nearly pseudo-Kähler manifolds.

Corollary 3.3.6 *The twistor spaces $\mathcal{Z}$ of*

$$\tilde{\mathbb{H}}P^1 = \mathrm{Sp}(1,1)/\mathrm{Sp}(1)\mathrm{Sp}(1) \, and \, SU(1,2)/S(U(1)U(2))$$

provide six-dimensional nearly pseudo-Kähler manifolds.

Remark 3.3.7 The situation in negative scalar curvature is more flexible than in the positive case. This is illustrated by the following results in this area: In the main theorem of [89] it is shown that the moduli space of complete quaternionic Kähler metrics on $\mathbb{R}^{4n}$ is infinite dimensional. A construction of super-string theory, called the *c-map* [54], yields continuous families of negatively curved quaternionic Kähler manifolds. Let us mention, that the c-map enjoys very recent interest [10, 73, 93] in differential geometry. These results show that Corollary 3.3.4 is a good source of examples.

Another source of examples is given by twistor spaces over *para-quaternionic Kähler manifolds*. Since these manifolds are less classical than quaternionic Kähler manifolds, we recall some definitions (cf. [5] and references therein).

Definition 3.3.8 Let $(\kappa_1, \kappa_2, \kappa_3) = (-1, 1, 1)$ or some permutation thereof. An almost para-quaternionic structure on a differentiable manifold M^{4k} is a rank 3 sub-bundle $Q \subset End(TM)$, which is locally generated by three anti-commuting endomorphism-fields $J_1, J_2, J_3 = J_1 J_2$. These satisfy $J_i^2 = \kappa_i Id$ for $i = 1, \ldots, 3$. Such a triple is called standard local basis of Q. A linear torsion-free connection preserving Q is called para-quaternionic connection. An almost para-quaternionic structure is called a para-quaternionic structure if it admits a para-quaternionic connection. An almost para-quaternionic Hermitian structure (M, Q, g) is a pseudo-Riemannian manifold endowed with a para-quaternionic structure such that Q consists of skew-symmetric endomorphisms. For $n > 1$ (M^{4k}, Q, g) is a para-quaternionic Kähler manifold if Q is preserved by the Levi-Civita connection of g. In dimension 4 a para-quaternionic Kähler manifold M^4 is an anti-self-dual Einstein manifold.

We use the same notions omitting the word *para* for the quaternionic case. The condition that Q is preserved by the Levi-Civita connection is in a given standard local basis $\{J_i\}_{i=1}^3$ of Q equivalent to the equations

$$\nabla_X J_i = -\theta_k(X)\kappa_j J_j + \theta_j(X)\kappa_k J_k, \text{ for } X \in TM, \tag{3.40}$$

where i, j, k is a cyclic permutation of $1, 2, 3$ and $\{\theta_i\}_{i=1}^3$ are local one-forms. In the context of para-quaternionic manifolds one can define twistor spaces for $s = 1, 0, -1$

$$\mathcal{Z}^s := \{A \in Q \,|\, A^2 = sId, \text{ with } A \neq 0\}.$$

The case of interest in this text is $\mathcal{Z} = \mathcal{Z}^{-1}$, since this twistor space is a complex manifold, such that the conditions of Proposition 3.3.2 hold true (cf. [5]). Therefore we obtain the following examples of nearly pseudo-Kähler manifolds.

Corollary 3.3.9 *The twistor space $\mathcal{Z}$ of a para-quaternionic Kähler manifold with non-zero scalar curvature of dimension $4k$ admits a canonical nearly pseudo-Kähler structure of reducible holonomy contained in $U(k, k) \times U(1)$.*

Example 3.3.10 The para-quaternions $\widetilde{\mathbb{H}}$ are the $\mathbb{R}$-algebra generated by $\{1, i, j, k\}$ subject to the relations $i^2 = -1$, $j^2 = k^2 = 1$, $ij = -ji = k$. Like the quaternions, the para-quaternions are a real Clifford algebra which in the convention of [88] is $\widetilde{\mathbb{H}} = Cl_{1,1} \cong Cl_{0,2} \cong \mathbb{R}(2)$. One defines the para-quaternionic projective space $\widetilde{\mathbb{H}}P^n$ by the obvious equivalence relation on the para-quaternionic right-module $\widetilde{\mathbb{H}}^{n+1}$ of $(n+1)$-tuples of para-quaternions. The manifold $\widetilde{\mathbb{H}}P^n$ is a para-quaternionic Kähler manifold [21] in analogue to the quaternionic projective space $\mathbb{H}P^n$. This yields examples of the type described in the last Corollary.

3.4   Complex Reducible Nearly Pseudo-Kähler Manifolds

Motivation In this section we study the case of a nearly pseudo-Kähler manifold (M^{2n}, J, g), such that the holonomy of the characteristic connection $\bar{\nabla}$ is reducible, in the sense that the tangent bundle TM admits a splitting

$$TM = \mathcal{H} \oplus \mathcal{V}$$

into two $\bar{\nabla}$-parallel sub-bundles $\mathcal{H}, \mathcal{V}$, which are orthogonal and invariant with respect to the almost complex structure J. We refer to this situation as complex reducible. This is motivated by the examples on twistor spaces given in the last section. In Sect. 3.9.4 we see, that *real reducible* nearly Kähler manifolds are locally homogeneous.

3.4.1   General Properties

In this subsection we carefully check, generalising [99] to pseudo-Riemannian foliations, the information which follows from the decomposition into the J-invariant sub-bundles.

Lemma 3.4.1 (Lemma 6.1 of [108]) *In the situation of this section and for a vector field X in $\mathcal{H}$, a vector field Y in TM and vector fields U, V in $\mathcal{V}$ it is*

$$\bar{R}(X, Y, U, V) = g\left([\nabla_U J, \nabla_V J]X, Y\right) - g\left((\nabla_X J)Y, (\nabla_U J)V\right). \qquad (3.41)$$

Corollary 3.4.2 *For vector fields X, Y in $\mathcal{H}$ and V, W in $\mathcal{V}$ one has*

  (i) $(\nabla_X J)(\nabla_V J)W = 0$; $(\nabla_V J)(\nabla_X J)Y = 0$;
 (ii) $(\nabla_X J)(\nabla_Y J)Z$ belongs to $\mathcal{H}$ for all $Z \in \Gamma(\mathcal{H})$;
(iii) $(\nabla_V J)(\nabla_W J)X$ belongs to $\mathcal{H}$; and $(\nabla_X J)(\nabla_Y J)V$ belongs to $\mathcal{V}$.

Proof

  (i) follows from the fact, that $\bar{R}(JX, JY, V, W) = \bar{R}(X, Y, V, W)$ and that the first term of Eq. (3.41) has the same symmetry with respect to J. This yields on the one hand

$$g\left((\nabla_{JX} J)JY, (\nabla_V J)W\right) = g\left((\nabla_X J)Y, (\nabla_V J)W\right)$$

and on the other hand it is

$$g\left((\nabla_{JX} J)JY, (\nabla_V J)W\right) = -g\left((\nabla_X J)Y, (\nabla_V J)W\right).$$

Consequently one has $g\left((\nabla_X J)Y, (\nabla_V J)W\right) = 0$. Exchanging $\mathcal{H}$ and $\mathcal{V}$ finishes part (i).

(ii) From (i) one gets the vanishing of

$$g((\nabla_V J)(\nabla_Y J)Z, X) = g(Z, (\nabla_Y J)(\nabla_V J)X)$$
$$= -g(Z, (\nabla_Y J)(\nabla_X J)V) = -g((\nabla_X J)(\nabla_Y J)Z, V).$$

(iii) From (i) it follows $0 = \bar{R}(X, U, V, W) = g([\nabla_V J, \nabla_W J]X, U)$. This yields $[\nabla_V J, \nabla_W J]X \in \mathcal{H}$ and by $[\nabla_V J, \nabla_{JW} J]JX = -\{\nabla_V J, \nabla_W J\}X \in \mathcal{H}$ we get the first part. The second part follows by replacing $\mathcal{H}$ and $\mathcal{V}$.

$\square$

3.4.2 Co-dimension Two

Motivated by the above section on twistor spaces we suppose from now on that the real dimension of $\mathcal{V}$ is two.

Lemma 3.4.3 (Lemma 6.2 of [108]) *Let* $\dim_{\mathbb{R}}(\mathcal{V}) = 2$.

(i) *Then the restriction of the metric g is either of signature* $(2, 0)$ *or* $(0, 2)$.
(ii) a) $T(V, W) = 0$ *for all* $V, W \in \mathcal{V}$.
   b) $T(X, U) \in \mathcal{H}$ *for all* $X \in \mathcal{H}$ *and* $U \in \mathcal{V}$.
   c) *In dimension 6 it is* $T(X, Y) \in \mathcal{V}$ *for all* $X, Y \in \mathcal{H}$.
   d) $\mathrm{Span}\{\pi_{\mathcal{V}}(T(X, Y)) \mid X, Y \in \mathcal{H}\} = \mathcal{V}$.

Corollary 3.4.4 *Let* $\dim_{\mathbb{R}}(\mathcal{V}) = 2$. *Then the foliation* $\mathcal{V}$ *has totally geodesic fibres and the O'Neill tensor is given by* $A_X Y = \frac{1}{2}\pi_{\mathcal{V}}(J(\nabla_X J)Y)$ *and* $A_X V = \frac{1}{2}J(\nabla_X J)V$. *Moreover it is* $\nabla^{\mathcal{V}} J = 0$.

Proof From Lemma 3.4.3 (ii) a) we obtain $(\nabla_V J)W = 0$ with $V, W \in \Gamma(\mathcal{V})$. By Lemma 3.3.3 part 2) it follows $T_V W = 0$ and $\nabla^{\mathcal{V}} J = 0$, since the decomposition $\mathcal{H} \oplus \mathcal{V}$ is $\bar{\nabla}$ parallel. Part 1) of Lemma 3.3.3 finishes the proof. $\square$

Proposition 3.4.5 *Let* (M, J, g) *be a nearly pseudo-Kähler manifold such that the property of Lemma 3.4.3 (ii) c) is satisfied and such that* $\mathcal{V}$ *has dimension 2, then* $(M, \check{J} = \hat{J}, \check{g} = g_2)$ *is a pseudo-Kähler manifold.*[4]
It is natural to suppose the property of Lemma 3.4.3 (ii) c), since this holds true in the cases of *twistorial type* which are studied in the next sections.

Proof By the last Corollary the data of the submersion is $\check{T} = T \equiv 0$, $A_X Y = \check{A}_X Y = \frac{1}{2}\pi_{\mathcal{V}}(J(\nabla_X J)Y)$ and $\check{A}_X V = 2A_X V = J(\nabla_X J)V$. Since A anti-commutes with J it commutes with $\check{J}$. This yields the conditions (3.33) and (3.34) of

---

[4]Here we use $\check{}$ for the inverse construction of $\hat{}$.

Lemma 3.3.1 on the triple $\check{A}, \check{T}, \check{J}$. Further it holds $\nabla^{\mathcal{V}}J = 0$. From the reasoning of Eq. (3.35) we obtain $\pi_{\mathcal{H}}((\check{\nabla}_X\check{J})Y) = \pi_{\mathcal{H}}((\nabla_XJ)Y)$ which vanishes by the property of Lemma 3.4.3 (ii) c). By an analogous argument we get from Eq. (3.36) the identity $\pi_{\mathcal{V}}((\check{\nabla}_X\check{J})V) = -\pi_{\mathcal{V}}((\nabla_XJ)V)$. This vanishes by Lemma 3.4.3 (ii) b). From Eq. (3.37) we derive $-\pi_{\mathcal{H}}((\nabla_VJ)V) \overset{n.K.}{=} \pi_{\mathcal{H}}((\nabla_VJ)X) = \pi_{\mathcal{H}}((\check{\nabla}_V\check{J})X) + 2\pi_{\mathcal{H}}(JA_XV)$. The definition of A_XV yields $\pi_{\mathcal{H}}((\check{\nabla}_V\check{J})X) = 0$. These are all the identities needed to apply Lemma 3.3.1.   □

Proposition 3.4.6 *Let X, Y be vector fields in $\mathcal{H}$ and V_1, V_2, V_3 be vector fields in $\mathcal{V}$. Suppose that it holds $T(V, W) = 0$ for all $V, W \in \mathcal{V}$ then it is*

$$\bar{R}((\nabla_XJ)JY, V_1, V_2, V_3) = g(JY, [\nabla_{V_1}J, [\nabla_{V_2}J, \nabla_{V_3}J]]X). \tag{3.42}$$

Moreover, one has $\bar{\nabla}_U\bar{R}(V_1, V_2, V_3, V_4) = 0$.

Proof For $V_1, V_2, V_3 \in \mathcal{V}$ and $X \in \mathcal{H}$ the second Bianchi identity gives

$$- \underset{XYV_1}{\sigma}\, \bar{\nabla}_X(\bar{R})(Y, V_1, V_2, V_3) = \underset{XYV_1}{\sigma}\, \bar{R}((\nabla_XJ)JY, V_1, V_2, V_3).$$

As the decomposition $\mathcal{H} \oplus \mathcal{V}$ is $\bar{\nabla}$-parallel the terms on the left hand-side vanish due to the symmetries (3.24) of the curvature tensor $\bar{R}$. The right hand-side is determined with the help of Lemma 3.4.1 and Corollary 3.4.2. If we apply $\bar{\nabla}$ to the formula (3.42) we obtain by $\bar{\nabla}(\nabla J) = 0$ the identity $g(\bar{\nabla}_U(\bar{R})(V_1, V_2, V_3), (\nabla_XJ)Z) = 0$ with $Z = JY$. This yields the proposition using Lemma 3.4.3 (ii) part d).   □

3.4.3 Six-Dimensional Nearly Pseudo-Kähler Manifolds

Before analysing the general case we first focus on dimension 6.

Lemma 3.4.7 (Lemma 6.7 of [108]) *On a six-dimensional nearly pseudo-Kähler manifold (M^6, J, g) the integral manifolds of the foliation $\mathcal{V}$ have Gaussian curvature 4α and constant curvature $\kappa = 4\alpha$, where α is the type constant.*

Let us recall that the sign of α is completely determined by the signature of the metric g, cf. Remark 3.1.9.

Proposition 3.4.8 *The manifold (M, J, g) is the total space of a pseudo-Riemannian submersion $\pi : (M, g) \to (N, h)$ where (N, h) is an almost pseudo-Hermitian manifold and the fibres are totally geodesic Hermitian symmetric spaces. In particular, the fibres are simply connected.*

Proof The foliation which is induced by $\mathcal{V}$ is totally geodesic and each leaf is by Proposition 3.4.6 a locally Hermitian symmetric space of complex dimension 1.

It is shown in Lemma 3.4.7 that each leaf has constant curvature κ. In the case $\kappa > 0$ the leaves are compact and we can apply a result of Kobayashi, cf. [18,

11.26], to obtain that the leaves are simply connected. Since the leaves are also simply connected it follows, that the leaf holonomy is trivial and that the foliation comes from a (smooth) submersion (cf. p. 90 of [113]). In the case $\kappa < 0$ we observe, that $(M, J, -g)$ is a nearly pseudo-Kähler manifold of constant type $-\alpha$. The same argument shows that the fibres are simply connected. $\qquad\square$

Lemma 3.4.9 (Lemma 6.9 of [108]) *Let (M^6, g, J) be a strict nearly pseudo-Kähler six-manifold of constant type α. For an arbitrary normalised[5] local vector field $V \in \mathcal{V}$, i.e. $\epsilon_V = g(V, V) \in \{\pm 1\}$, we consider the endomorphisms $\tilde{J}_1 := J_{|\mathcal{H}}$, $\tilde{J}_2 : \mathcal{H} \ni X \mapsto \sqrt{|\alpha|}^{-1}(\nabla_V J)X \in \mathcal{H}$ and $\tilde{J}_3 = \tilde{J}_1\tilde{J}_2$. Then the triple $(\tilde{J}_1, \tilde{J}_2, \tilde{J}_3)$ defines an ε-quaternionic triple on $\mathcal{H}$ with $\kappa_1 = -1$ and $\kappa_2 = \kappa_3 = sign(-\alpha\epsilon_V)$ and it is*

$$\pi^{\mathcal{H}}[(\nabla_\chi \tilde{J}_i)Y] = -\theta_k(\chi)\,\kappa_j\tilde{J}_jY + \theta_j(\chi)\,\kappa_k\tilde{J}_kY,$$

for a cyclic perm. of i,j,k and with $\theta_1(\chi) = sign(\alpha)g(JV, \bar\nabla_\chi V)$, $\theta_2(\chi) = -sign(\alpha)\sqrt{|\alpha|}g(V, J\chi)$ and $\theta_3(\chi) = sign(\alpha)\sqrt{|\alpha|}g(V, \chi)$. The sub-bundle of endomorphisms spanned by $(\tilde{J}_1, \tilde{J}_2, \tilde{J}_3)$ does not depend on the choice of V.

Lemma 3.4.10 *Let (M^6, g, J) be a strict nearly pseudo-Kähler six-manifold of constant type α. Let $s : U \subset N \to M$ be a (local) section[6] of π on some open set U. Define ϕ by*

$$\phi = s_* \circ \pi_* : \mathcal{H}_{s(n)} \overset{\pi_*}{\to} T_n N \overset{s_*}{\to} s_*(T_n N) \subset T_{s(n)}M, \text{ for } n \in N$$

and set $J_{i|n} := \pi_* \circ \tilde{J}_{i|s(n)} \circ (\pi_{*|\mathcal{H}})^{-1}$ for $i = 1, \ldots, 3$, where $\tilde{J}_i$ are defined in Lemma 3.4.9. Then (J_1, J_2, J_3) defines a local ϵ-quaternionic basis preserved by the Levi-Civita connection ∇^N of N.

Proof We choose U such that the section s is a diffeomorphism onto $W = s(U)$ and a vector field V in $\mathcal{V}$ defined on a subset containing W. As π is a pseudo-Riemannian submersion we obtain from $\pi_* \circ s_* = Id$ that s is an isometry from U onto W. Therefore it holds $s_*(\nabla_X^N Y) = \pi^{s_*TN}[\nabla_{s_*X}s_*Y]$ which yields $\nabla_X^N Y = \pi_*(\nabla_{s_*X}s_*Y)$ and

$$(\pi_{*|\mathcal{H}})^{-1}(\nabla_X^N Y) = \pi^{\mathcal{H}}(\nabla_{s_*X}s_*Y). \tag{3.43}$$

For convenience let us identify U and W or in other words consider s as the inclusion $W \subset M$. Then the projection on s_*TN is $\phi = s_*\pi_* = \pi_*|\mathcal{H}$. Moreover, we need the (tensorial) relation[7]

$$\nabla_X^N(\pi_* Z) - \pi_*\pi^{\mathcal{H}}(\nabla_X^M Z) = 0 \text{ or equivalently } \nabla_X^N \tilde{Z} - \pi_*\pi^{\mathcal{H}}(\nabla_X^M \phi^{-1}\tilde{Z}) = 0,$$

---

[5]Constant non-zero length suffices.

[6]Local sections exist, since π is locally trivial [18, 9.3].

[7]Here $\tilde{Z}$ is the horizontal lift of Z.

which can be directly checked for basic vector fields. Using this identity we get for $i = 1, \ldots, 3$

$$\nabla_X^N(J_i Y) = \nabla_X^N(\phi \tilde{J}_i \phi^{-1} Y) = \phi \nabla_X^M(\tilde{J}_i \phi^{-1} Y) = \phi (\nabla_X^M \tilde{J}_i) \phi^{-1} Y + \phi \tilde{J}_i \nabla_X^M(\phi^{-1} Y)$$
$$= \phi (\nabla_X^M \tilde{J}_i) \phi^{-1} Y + \phi \tilde{J}_i \phi^{-1} \nabla_X^N Y = \phi (\nabla_X^M \tilde{J}_i) \phi^{-1} Y + J_i \nabla_X^N Y,$$

which reads $(\nabla_X^N J_i) Y = \phi (\nabla_X^M \tilde{J}_i) \phi^{-1} Y$. This finishes the proof, since the right hand-side is completely determined by Lemma 3.4.9. Therefore we have checked the condition (3.40), i.e. the manifold N is endowed with a parallel skew-symmetric (para-)quaternionic structure, see also [18, 10.32 and 14.36].   □

3.4.4   General Dimension

In the last section we have seen that in dimension 6 the tensor $\nabla_V J$ induces a (para-)complex structure on $\mathcal{H}$. This motivates the following definition.

Definition 3.4.11 The foliation induced by $TM = \mathcal{H} \oplus \mathcal{V}$ is called of twistorial type if for all $p \in M$ there exists a $V \in \mathcal{V}_p$ such that the endomorphism

$$\nabla_V J : \mathcal{H}_p \to \mathcal{H}_p$$

is injective.

Obviously, if $\nabla_V J$ defines a (para-)complex structure, then the foliation is of twistorial type.

Proposition 3.4.12

(a) *If the metric induced on $\mathcal{H}$ is definite, then the foliation is of twistorial type.*
(b) *If the foliation is of twistorial type, then for all $p \in M$ and all $0 \neq U \in \mathcal{V}_p$ the endomorphism*

$$\nabla_U J : \mathcal{H}_p \to \mathcal{H}_p$$

is injective.
(c) *It holds with $A := \nabla_V J$ for some vector field V in $\mathcal{V}$ of constant length and for vector fields $X \in \mathcal{H}$ and $\chi \in TM$*

$$\bar{\nabla}_\chi(A^2) X = 0. \tag{3.44}$$

Further it holds $[A^2, (\nabla_U J)] = 0$ for all $U \in \mathcal{V}$ and

$$\nabla_U(A^2) X = 0 \tag{3.45}$$

for vector fields U in $\mathcal{V}$.

Proof Part (a) follows from $(\nabla_V J)X \in \mathcal{H}$ for $X \in \mathcal{H}$ and $V \in \mathcal{V}$, cf. Lemma 3.4.3 (i). For (b) we observe, that if $\nabla_V J$ is injective so is $\nabla_{JV} J = -J\nabla_V J$. As $\mathcal{V}$ is of dimension 2 $\{V, JV\}$ with $V \neq 0$ is an orthogonal basis. With $a, b \in \mathbb{R}$ it follows $g((a\nabla_V J + b\nabla_{JV} J)X, (a\nabla_V J + b\nabla_{JV} J)X) = (a^2 + b^2)\,g((\nabla_V J)X, (\nabla_V J)X)$, which yields, that $\nabla_{aV+bJV} J : \mathcal{H}_p \to \mathcal{H}_p$ is injective since $a \neq 0$ or $b \neq 0$. It remains to prove part (c). We first observe, that, since V has constant length and since $\bar{\nabla}$ is a metric connection and preserves $\mathcal{V}$, it follows $\bar{\nabla}_X V = \alpha(\chi)JV$ for some one-form α. From $\bar{\nabla}(\nabla J) = 0$ we obtain

$$(\bar{\nabla}_\chi A)X = (\bar{\nabla}_\chi(\nabla_V J))X = (\nabla_{\bar{\nabla}_\chi V} J)X = \alpha(\chi)(\nabla_{JV} J)X = -\alpha(\chi)JAX$$

and we compute using $\{A, J\} = 0$

$$\bar{\nabla}_\chi(A^2)X = A(\bar{\nabla}_\chi A)X + (\bar{\nabla}_\chi A)AX = -\alpha(\chi)[A(J(AX)) + JA^2X] = 0.$$

The equation $[A^2, (\nabla_U J)] = 0$ is tensorial in U and holds true for $U = V$. Therefore we only need to compute $[A^2, (\nabla_{JV} J)] = -[A^2, J(\nabla_V J)] = -J[A^2, (\nabla_V J)] = 0$, where we used that A^2 commutes with J. This implies

$$\nabla_U(A^2)X = \bar{\nabla}_U(A^2)X + \frac{1}{2}[J(\nabla_U J), A^2]X = -\frac{1}{2}[(\nabla_{JU} J), A^2]X = 0$$

and proves part (c). $\qquad\square$

In the following V is a local vector field of constant length $\epsilon_V = g(V, V) \in \{\pm 1\}$.

We denote by Ω the curvature form of the connection induced by $\bar{\nabla}$ on the (complex) line bundle $\mathcal{V}$, which is given by

$$\bar{R}(X, Y)V = \Omega(X, Y)JV, \text{ for } X, Y \in TM, V \in \mathcal{V}.$$

Proposition 3.4.13 *If the foliation is of twistorial type,*

(i) *then the endomorphism $A := \nabla_V J|_{\mathcal{H}}$ satisfies $A^2 = \kappa\epsilon_V Id_{\mathcal{H}}$ for some real constant $\kappa \neq 0$ and*

$$\Omega = -2\kappa(2\omega^{\mathcal{V}} - \omega^{\mathcal{H}}),$$

where $\omega^{\mathcal{H}}(X, Y) = g(X, JY)$ is the restriction of the fundamental two-form ω to $\mathcal{H}$;

(ii) *for X, Y in $\mathcal{H}$ it is $(\nabla_X J)Y \in \mathcal{V}$.*

The proof of this proposition is divided in several steps.

Lemma 3.4.14

(i) *For X, Y in $\mathcal{H}$ and V in $\mathcal{V}$ it is $\bar{R}(X, Y, V, JV) = -2g((\nabla_V J)^2 X, JY)$.*

(ii) *For a given X in $\mathcal{H}$ and V in $\mathcal{V}$ it follows $\bar{R}(X, V, V, JV) = 0$.*

Proof

(i) Since $\mathcal{H}$ is $\bar{\nabla}$-parallel we obtain, that $\underset{XYV}{\sigma}\,\bar{R}(X,Y,V,JV) = \bar{R}(X,Y,V,JV)$. This is the left hand-side of the first Bianchi identity (3.26) . The right hand-side reads

$$-\underset{XYV}{\sigma}g((\nabla_XJ)Y,(\nabla_VJ)JV) = -g((\nabla_VJ)X,(\nabla_YJ)JV) - g((\nabla_YJ)V,(\nabla_XJ)JV)$$

$$= -2g((\nabla_VJ)^2X,JY).$$

(ii) From the symmetries (3.24) of the curvature tensor $\bar{R}$ it follows $\bar{R}(X,V,V,JV) = \bar{R}(V,JV,X,V)$. This expression vanishes since $\mathcal{H}$ is $\bar{\nabla}$-parallel.

$\square$

From the last lemma we derive the more explicit expression of the curvature form $\Omega(\cdot,\cdot)$:

$$\Omega(\cdot,\cdot) = f\omega^V(\cdot,\cdot) + \epsilon_V\alpha(\cdot,\cdot), \tag{3.46}$$

where f is a smooth function, ω^V is the restriction of the fundamental two-form $\omega = g(\cdot,J\cdot)$ to V and $\alpha(X,Y) = -2g(A^2X,JY)$.

Lemma 3.4.15 (Lemma 6.15 of [108]) *It holds with $U \in V$ and $X,Y \in \mathcal{H}$:*

$$d\omega^V(X,U,JU) = 0, \tag{3.47}$$

$$d\alpha(X,U,JU) = 0, \tag{3.48}$$

$$d\omega^V(U,X,Y) = -g((\nabla_UJ)X,Y), \tag{3.49}$$

$$d\alpha(U,X,Y) = 4g(A^2(\nabla_UJ)X,Y). \tag{3.50}$$

Proof of the Proposition 3.4.13

(i) Let X,Y be vector fields in $\mathcal{H}$ and V be a local vector field in V of constant length. Since Ω as a curvature form of a (Hermitian) line bundle is closed, we obtain from Eq. (3.46) $-\epsilon_V d\alpha(\cdot,\cdot,\cdot) = fd\omega^V(\cdot,\cdot,\cdot) + df \wedge \omega^V(\cdot,\cdot,\cdot)$. Equations (3.47) and (3.48) imply $df_{|\mathcal{H}} = 0$. This implies $[X,Y]f = 0$ and using that $\mathcal{H}$ is $\bar{\nabla}$-parallel we obtain $(\bar{\nabla}_XY)f = 0 = (\bar{\nabla}_YX)f$ which yields finally $0 = T(X,Y)(f) = -[J(\nabla_XJ)Y](f)$. By Lemma 3.4.3 (ii) d) the last equation shows $df_{|V} = 0$. Since M is connected, it follows $f \equiv -\kappa$ for a constant κ.

Again using $d\Omega(V,X,Y) = 0$ Eqs. (3.49) and (3.50) yield for arbitrary X,Y

$$\kappa g((\nabla_VJ)X,Y) + 4\epsilon_V g(A^2(\nabla_VJ)X,Y) = 0.$$

This implies $(\nabla_VJ)(\kappa Id_{\mathcal{H}} + 4\epsilon_V A^2) = 0$. It follows

$$A^2 = -\epsilon_V\frac{\kappa}{4}Id_{\mathcal{H}},$$

since the foliation is of twistorial type. If we set $4\alpha = \kappa$ in analogue to dimension 6^8 one gets $A^2 = -\epsilon_V \alpha Id_{\mathcal{H}}$.

(ii) Since Ω is closed, it follows from part (i) and $d\omega^V(X, Y, Z) = 0$ for $X, Y, Z \in \mathcal{H}$ that it is $d\omega^{\mathcal{H}}(X, Y, Z) = 0$. Using $d\omega^{\mathcal{H}}(X, Y, Z) = 3g((\nabla_X J)Y, Z)$ yields part (ii).

$\square$

Proposition 3.4.16 *Let (M^{4k+2}, g, J) be a strict nearly pseudo-Kähler manifold of twistorial type. Let $s : U \subset N \to M$ be a (local) section of π on some open set U. Define ϕ by*

$$\phi = s_* \circ \pi_* : \mathcal{H}_{s(n)} \overset{\pi_*}{\to} T_n N \overset{s_*}{\to} s_*(T_n N) \subset T_{s(n)}M, \text{ for } n \in N$$

and set $J_{i|n} := \pi_* \circ \tilde{J}_{i|s(n)} \circ (\pi_{*|\mathcal{H}})^{-1}$ for $i = 1, \ldots, 3$, where $\tilde{J}_i$ are defined in Lemma 3.4.9. Then (J_1, J_2, J_3) defines a local ϵ-quaternionic basis preserved by the Levi-Civita connection ∇^N of N.

Proof The proof of Lemma 3.4.9 only uses $A^2 = -\epsilon_V \frac{\kappa}{4} Id$ and $(\nabla_X J)Y \in \mathcal{V}$ for $X, Y \in \mathcal{H}$. Therefore we can generalise it by means of Proposition 3.4.13 to strict nearly pseudo-Kähler manifolds of twistorial type. $\square$

Corollary 3.4.17

(i) *The tensor r has exactly two eigenvalues. More precisely, it has the eigenvalue κ on $\mathcal{H}$ and the eigenvalue $\epsilon_V \frac{\kappa}{2}(n - 1)$ on $\mathcal{V}$ with $\kappa = 4\alpha$.*

(ii) *The Ricci-tensor has exactly two eigenvalues. More precisely, it has the eigenvalue $\frac{\kappa}{4}(\epsilon_V(n-1)+3)$ on $\mathcal{H}$ and the eigenvalue $\kappa\left(\epsilon_V \frac{(n-1)}{8} + 1\right)$ on $\mathcal{V}$. The base manifold (N, h) is an Einstein manifold with Einstein constant $\frac{\kappa}{4}\epsilon_V(n-1)$.*

Proof By definition we have

$$g(rX, Y) = \sum_{i=1}^{2n} \epsilon_i g((\nabla_X J)e_i, (\nabla_Y J)e_i) = -\sum_{i=1}^{2n} \epsilon_i g((\nabla_Y J)(\nabla_X J)e_i, e_i)$$

for some pseudo-orthogonal basis with $e_1, \ldots, e_{2n-2} \in \mathcal{H}$ and $e_{2n-1}, e_{2n} \in \mathcal{V}$. For $V \in \mathcal{V}$ with $g(V, V) = \epsilon_V$ we get

$$g(rV, V) = -\sum_{i=1}^{2n} \epsilon_i g((\nabla_V J)(\nabla_V J)e_i, e_i) = -\sum_{i=1}^{2n-2} \epsilon_i g(A^2 e_i, e_i)$$

$$= \epsilon_V \frac{\kappa}{4} \sum_{i=1}^{2n-2} \epsilon_i g(e_i, e_i) = \epsilon_V \frac{\kappa}{2}(n - 1).$$

---

[8]Without risk of confusion we use the same latter for the constant α as for the two-form $\alpha(\cdot, \cdot)$.

Let us now consider $X \in \mathcal{H}$ and V as before and compute

$$g(rX, X) = 2\epsilon_V g\left((\nabla_V J)X, (\nabla_V J)X\right) + \sum_{i=1}^{2n-2} \epsilon_i g((\nabla_X J)e_i, (\nabla_X J)e_i).$$

Since it is $(\nabla_X J)e_i \in \mathcal{V}$, we get

$$\epsilon_i g((\nabla_X J)e_i, (\nabla_X J)e_i) = \epsilon_i \epsilon_V (g((\nabla_X J)e_i, V)^2 + g((\nabla_X J)e_i, JV)^2)$$

and for the sum this gives

$$g(r^{\mathcal{H}}X, X) = \sum_{i=1}^{2n-2} \epsilon_i g((\nabla_X J)e_i, (\nabla_X J)e_i)$$

$$= \epsilon_V \sum_{i=1}^{2n-2} \epsilon_i \left(g((\nabla_X J)V, e_i)^2 + g((\nabla_X J)JV, e_i)^2\right)$$

$$= 2\epsilon_V g\left((\nabla_V J)X, (\nabla_V J)X\right) = \frac{\kappa}{2} g(X, X).$$

Summarizing it follows $g(rX, X) = 4\epsilon_V g\left((\nabla_V J)X, (\nabla_V J)X\right) = \kappa$. This shows part (i).

The statement (ii) follows from (i) using Lemma 3.2.7. Namely, for $X, Y \in \mathcal{H}$ it is

$$g(Ric(X), Y) = \frac{\kappa}{4} g(X, Y) + \epsilon_V \frac{1}{\kappa} \frac{\kappa}{2}(n-1) \underbrace{g(r^{\mathcal{V}}X, Y)}_{\frac{\kappa}{2} g(X,Y)} + g(r^{\mathcal{H}}X, Y) = \frac{\kappa}{4}(\epsilon_V(n-1)+3)g(X, Y),$$

since it is using $A^2 = -\epsilon_V \frac{\kappa}{4} Id_{\mathcal{H}}$

$$g(r^{\mathcal{V}}X, Y) = -\mathrm{tr}_{\mathcal{V}}\left((\nabla_X J) \circ (\nabla_Y J)\right) = -2\epsilon_V g((\nabla_X J)(\nabla_Y J)V, V)$$

$$= 2\epsilon_V g((\nabla_Y J)V, (\nabla_X J)V) = 2\epsilon_V g((\nabla_V J)Y, (\nabla_V J)X)$$

$$= 2\epsilon_V g(AY, AX) = \frac{\kappa}{2} g(X, Y).$$

Further, for $U, V \in \mathcal{V}$ it is

$$g(Ric(U), V) = \epsilon_V \frac{\kappa(n-1)}{8} g(U, V) + \epsilon_V \frac{2}{\kappa(n-1)} \kappa \underbrace{g(r^{\mathcal{H}}U, V)}_{\epsilon_V \frac{\kappa}{2}(n-1)g(U,V)}$$

$$= \kappa \left(\epsilon_V \frac{(n-1)}{8} + 1\right) g(U, V),$$

where

$$g(r^{\mathcal{H}}V, V) = -\operatorname{tr}_{\mathcal{H}}\left((\nabla_V J) \circ (\nabla_V J)\right) = -\sum_{i=1}^{2n-2} \epsilon_i \, g((\nabla_V J)(\nabla_V J)e_i, e_i)$$

$$= 2(n-1)\frac{\kappa}{4}\epsilon_V = \epsilon_V \frac{\kappa}{2}(n-1).$$

The last statement follows from O'Neill's formula with the information, that the O'Neill tensor is $A_X Y = \frac{1}{2}J(\nabla_X J)Y$, c.f. Lemma 3.3.3. □

3.4.5 The Twistor Structure

In this subsection we finally characterise the nearly pseudo-Kähler structures, which are related to the canonical nearly Kähler structure of twistor spaces.

Theorem 3.4.18 *Suppose, that (M^{2n}, J, g) is a complex reducible nearly pseudo-Kähler manifold of twistorial type, then one has:*

(i) The manifold $(M, J = \check{J}, \check{g} = g_2)$ is a twistor space of a quaternionic pseudo-Kähler manifold, if it is $\epsilon_V \kappa > 0$.

(ii) The manifold $(M, J = \check{J}, \check{g} = g_2)$ is a twistor space of a para-quaternionic Kähler manifold, if it is $\epsilon_V \kappa < 0$.

Proof Denote by $\pi^{\mathcal{Z}} : \mathcal{Z} \to N$ the twistor space of the manifold N endowed with the parallel skew-symmetric (para-)quaternionic structure constructed from the foliation $\pi : M \to N$ of twistorial type, cf. Proposition 3.4.9 for dimension 6 and Proposition 3.4.16 for general dimension. We observe that the restriction of J to $\mathcal{H}$ yields a (smooth) map

$$\varphi : M \to \mathcal{Z}, \quad m \mapsto d\pi_m \circ J_m \mid_{\mathcal{H}} \circ (d\pi_m|_{\mathcal{H}})^{-1} =: j_{\pi(m)},$$

which by construction satisfies $\pi^{\mathcal{Z}} \circ \varphi = \pi$ and as a consequence $d\pi^{\mathcal{Z}} \circ d\varphi = d\pi$. Since π and $\pi^{\mathcal{Z}}$ are pseudo-Riemannian submersions, the last equation implies that $d\varphi$ induces an isometry of the according horizontal distributions and maps the vertical spaces into each other. Let us determine the differential of φ on $\mathcal{V}$.

Claim: For $V \in \mathcal{V}$ one has

$$d\varphi(V) = 2 \, d\pi \circ (\nabla_V J) \circ (d\pi_{|\mathcal{H}})^{-1},$$

$$d\varphi(JV) = 2 \, d\pi \circ (\nabla_{JV} J) \circ (d\pi_{|\mathcal{H}})^{-1} = -2 \, d\pi \circ J(\nabla_V J) \circ (d\pi_{|\mathcal{H}})^{-1}.$$

To prove the claim we consider a (local) vector field $V \in \mathcal{V}$ and a (local) integral curve γ of V on some interval $I \ni 0$ with $\gamma(0) = m$. Let X be a vector field in N. Denote by $\tilde{X}$ the horizontal lift of X. The Lie transport of $\tilde{X}$ along the vertical curve γ projects to X, i.e. it holds $d\pi_{\gamma(t)}(\tilde{X}) = X$ for all $t \in I$ and in consequence $\left(d\pi_{\gamma(t)}{}_{|\mathcal{H}}\right)^{-1} X = \tilde{X}$. In other words $d\pi$ commutes with this Lie transport, which implies

$$d\varphi(V)X = d\pi((\mathcal{L}_V J)\tilde{X}),$$

as one directly checks using basic vector fields. Therefore we need to determine the Lie-derivative $\mathcal{L}$ of J :

$$\pi^{\mathcal{H}}((\mathcal{L}_V J)\tilde{X}) = \pi^{\mathcal{H}}([V, J\tilde{X}] - J[V, \tilde{X}])$$
$$= \pi^{\mathcal{H}} \left(\nabla_V(J\tilde{X}) - \nabla_{J\tilde{X}} V - J\nabla_V \tilde{X} + J\nabla_{\tilde{X}} V \right)$$
$$= \pi^{\mathcal{H}} \left((\nabla_V J)\tilde{X} - \frac{1}{2} J(\nabla_{J\tilde{X}} J) V + \frac{1}{2} J \left(J(\nabla_{\tilde{X}} J) \right) V \right) = 2(\nabla_V J)\tilde{X}.$$

This shows $d\varphi(V) = 2\, d\pi \circ (\nabla_V J) \circ (d\pi_{|\mathcal{H}})^{-1}$, which implies $d\varphi(JV) = 2\, d\pi \circ (\nabla_{JV} J) \circ (d\pi_{|\mathcal{H}})^{-1} = -2\, d\pi \circ J(\nabla_V J) \circ (d\pi_{|\mathcal{H}})^{-1}$. Given a local section $s : N \to M$ and the associated adapted frame of the (para-)quaternionic structure it follows that $\varphi \circ s$ is J_1, $d\varphi(V)$ is related to J_2 and $d\varphi(JV)$ to $-J_3$ which span the tangent space of the fibre $F_{\pi(m)} = S^2$ in $\varphi(m)$. The complex structure of $\mathcal{Z}$ maps J_2 to J_3. Hence $d\varphi$ is complex linear for the opposite complex structure $\check{J}$ on M. Further one sees in this local frame that φ maps horizontal part into horizontal part. Therefore φ is an isometry for the metric $\check{g} = g_2$, where the parameter in the canonical variation of the metric g is $t = 2$. This means that $(M, \check{J}, \check{g} = g_2)$ is isometrically biholomorph to $\mathcal{Z}$. $\qquad\square$

Combining Theorems 3.2.10 and 3.4.18 we obtain the following result.

Theorem 3.4.19 *Let (M^{10}, J, g) be a nice decomposable nearly pseudo-Kähler manifold, then the universal cover of M is either the product of a pseudo-Kähler surface and a (strict) nearly pseudo-Kähler manifold M^6 or a twistor space of an eight-dimensional (para-)quaternionic Kähler manifold endowed with its canonical nearly pseudo-Kähler structure.*

3.5 A Class of Flat Pseudo-Riemannian Lie Groups

In this section we consider flat pseudo-Riemannian Lie groups which are closely related to nearly Kähler geometry (cf. Sect. 3.6). These geometric objects are also of independent interest [13]. Let $V = (\mathbb{R}^n, \langle \cdot, \cdot \rangle)$ be the standard pseudo-Euclidian vector space of signature (k, l), $n = k + l$. Using the (pseudo-Euclidian) scalar product we shall identify $V \cong V^*$ and $\Lambda^2 V \cong \mathfrak{so}(V)$. These identifications provide the inclusion $\Lambda^3 V \subset V^* \otimes \mathfrak{so}(V)$. Using it we consider a three-vector $\eta \in \Lambda^3 V$ as an $\mathfrak{so}(V)$-valued one-form. Further we denote by $\eta_X \in \mathfrak{so}(V)$ the evaluation of this one-form on a vector $X \in V$. Let us recall, that the support of $\eta \in \Lambda^3 V$ is defined by

$$\Sigma_\eta := \mathrm{span}\{\eta_X Y \,|\, X, Y \in V\} \subset V. \tag{3.51}$$

Theorem 3.5.1 *Each*

$$\eta \in \mathcal{C}(V) := \{\eta \in \Lambda^3 V \,|\, \Sigma_\eta \ (totally)\ isotropic\} = \bigcup_{L \subset V} \Lambda^3 L$$

defines a 2-step nilpotent simply transitive subgroup $\mathcal{L}(\eta) \subset Isom(V)$, where the union runs over all maximal isotropic subspaces. The subgroups $\mathcal{L}(\eta)$, $\mathcal{L}(\eta') \subset Isom(V)$ associated to $\eta, \eta' \in \mathcal{C}(V)$ are conjugated if and only if $\eta' = g \cdot \eta$ for some element of $g \in O(V)$.

Proof Using Lemma 2.3.2 of Chap. 2 any three-vector $\eta \in \Lambda^3 V$ satisfies $\eta \in \Lambda^3 \Sigma_\eta$. This implies the equation $\mathcal{C}(V) = \bigcup_{L \subset V} \Lambda^3 L$. Let an element $\eta \in \mathcal{C}(V)$ be given. By Lemma 2.3.4 of Chap. 2 its support Σ_η is isotropic if and only if the endomorphisms $\eta_X \in \mathfrak{so}(V)$ satisfy $\eta_X \circ \eta_Y = 0$ for all $X, Y \in V$. The 2-step nilpotent group

$$\mathcal{L}(\eta) := \left\{ \exp\begin{pmatrix} \eta_X X \\ 0 \ \ 0 \end{pmatrix} = \begin{pmatrix} Id + \eta_X & X \\ 0 & 1 \end{pmatrix} \,\middle|\, X \in V \right\}$$

acts simply transitively on $V \cong V \times \{1\} \subset V \times \mathbb{R}$ by isometries:

$$\begin{pmatrix} Id + \eta_X & X \\ 0 & 1 \end{pmatrix}\begin{pmatrix} 0 \\ 1 \end{pmatrix} = \begin{pmatrix} X \\ 1 \end{pmatrix}.$$

Let us consider next $\eta, \eta' \in \mathcal{C}(V)$, $g \in O(V)$. The computation

$$g\mathcal{L}(\eta)g^{-1} = \left\{ \begin{pmatrix} Id + g\eta_X g^{-1} & gX \\ 0 & 1 \end{pmatrix} \,\middle|\, X \in V \right\} = \left\{ \begin{pmatrix} Id + g\eta_{g^{-1}Y}g^{-1} & Y \\ 0 & 1 \end{pmatrix} \,\middle|\, Y \in V \right\}$$

shows that $g\mathcal{L}(\eta)g^{-1} = \mathcal{L}(\eta')$ if and only if $\eta'_X = (g \cdot \eta)_X = g\,\eta_{g^{-1}X}\,g^{-1}$ for all $X \in V$. $\square$

Let $\mathcal{L} \subset Isom(V)$ be a simply transitive group. Pulling back the scalar product on V by the orbit map $\mathcal{L} \ni g \mapsto g0 \in V$ yields a left-invariant flat pseudo-Riemannian metric h on $\mathcal{L}$. A pair $(\mathcal{L}, h)$ consisting of a Lie group $\mathcal{L}$ and a flat left-invariant pseudo-Riemannian metric h on $\mathcal{L}$ is called a **flat pseudo-Riemannian Lie group**.

Theorem 3.5.2

(i) *The class of flat pseudo-Riemannian Lie groups $(\mathcal{L}(\eta), h)$ defined in Theorem 3.5.1 exhausts all simply connected flat pseudo-Riemannian Lie groups with bi-invariant metric.*

(ii) *A Lie group with bi-invariant metric is flat if and only if it is 2-step nilpotent.*

Proof

(i) The group $\mathcal{L}(\eta)$ associated to a three-vector $\eta \in C(V)$ is diffeomorphic to $\mathbb{R}^n$ by the exponential map. We have to show that the flat pseudo-Riemannian metric h on $\mathcal{L}(\eta)$ is bi-invariant. The Lie algebra of $\mathcal{L}(\eta)$ is identified with the vector space V endowed with the Lie bracket

$$[X, Y] := \eta_X Y - \eta_Y X = 2\eta_X Y, \quad X, Y \in V.$$

The left-invariant metric h on $\mathcal{L}(\eta)$ corresponds to the scalar product $\langle \cdot, \cdot \rangle$ on V. Since $\eta \in \Lambda^3 V$, the endomorphisms $\eta_X = \frac{1}{2}ad_X$ are skew-symmetric. This shows that h is bi-invariant.

Conversely, let $(V, [\cdot, \cdot])$ be the Lie algebra of a pseudo-Riemannian Lie group of dimension n with bi-invariant metric h. We can assume that the bi-invariant metric corresponds to the standard scalar product $\langle \cdot, \cdot \rangle$ of signature (k, l) on V. Let us denote by $\eta_X \in \mathfrak{so}(V)$, $X \in V$, the skew-symmetric endomorphism of V which corresponds to the Levi-Civita covariant derivative D_X acting on left-invariant vector fields. From the bi-invariance and the Koszul formula we obtain that $\eta_X = \frac{1}{2}ad_X$ and, hence, $R(X, Y) = -\frac{1}{4}ad_{[X,Y]}$ for the curvature. The last formula shows that h is flat if and only if the Lie group is 2-step nilpotent. This proves (ii). To finish the proof of (i) we have to show that, under this assumption, η is completely skew-symmetric and has isotropic support. The complete skew-symmetry follows from $\eta_X = \frac{1}{2}ad_X$ and the bi-invariance. Similarly, using the bi-invariance, we have

$$4\langle \eta_X Y, \eta_Z W \rangle = \langle [X, Y], [Z, W] \rangle = -\langle Y, [X, [Z, W]] \rangle = 0,$$

since the Lie algebra is 2-step nilpotent. This shows that Σ_η is isotropic. $\square$

Corollary 3.5.3 *With the above notations, let $L \subset V$ be a maximally isotropic subspace. The correspondence $\eta \mapsto \mathcal{L}(\eta)$ defines a bijection between the points of the orbit space $\Lambda^3 L/GL(L)$ and isomorphism classes of pairs $(\mathcal{L}, h)$ consisting of a simply connected Lie group $\mathcal{L}$ endowed with a flat bi-invariant pseudo-Riemannian metric h of signature (k, l).*

Corollary 3.5.4 *Any simply connected Lie group $\mathcal{L}$ with a flat bi-invariant metric h of signature (k, l) contains a normal subgroup of dimension $\geq \max(k, l) \geq \frac{1}{2}\dim V$ which acts by translations on the pseudo-Riemannian manifold $(\mathcal{L}, h) \cong \mathbb{R}^{k,l}$.*

Proof Let $\mathfrak{a} := ker(X \mapsto \eta_X) \subset V$ be the kernel of η. Then $\mathfrak{a} = \Sigma_\eta^\perp$ is co-isotropic and defines an Abelian ideal $\mathfrak{a} \subset \mathfrak{l} := Lie \mathcal{L} \cong V \cong \mathbb{R}^{k,l}$. The corresponding normal subgroup $A \subset \mathcal{L} = \mathcal{L}(\eta)$ is precisely the subgroup of translations. So we have shown that $\dim A \geq \max(k, l) \geq \frac{1}{2} \dim V$. $\square$

Remarks 3.5.5

1) The number $\dim \Sigma_\eta$ is an isomorphism invariant of the groups $\mathcal{L} = \mathcal{L}(\eta)$, which is independent of the metric. We will denote it by $s(\mathcal{L})$. Let $L_3 \subset L_4 \subset \cdots \subset L$ be a filtration, where $\dim L_j = j$ runs from 3 to $\dim L$. The invariant $\dim \Sigma_\eta$ defines a decomposition of $\Lambda^3 L / GL(L)$ as a union

$$\{0\} \cup \bigcup_{j=3}^{\dim L} \Lambda_{reg}^3 L_j / GL(L_j),$$

where $\Lambda_{reg}^3 \mathbb{R}^j \subset \Lambda^3 \mathbb{R}^j$ is the open subset of 3-vectors with j-dimensional support. The points of the stratum $\Lambda_{reg}^3 L_j / GL(L_j) \cong \Lambda_{reg}^3 \mathbb{R}^j / GL(j)$ correspond to isomorphism classes of pairs $(\mathcal{L}, h)$ with $s(\mathcal{L}) = j$.
2) Since in the above classification Σ_η is isotropic, it is clear that a flat (or 2-step nilpotent) bi-invariant metric on a Lie group is indefinite, unless $\eta = 0$ and the group is Abelian. It follows from Milnor's classification of Lie groups with a flat left-invariant Riemannian metric [95] that a 2-step nilpotent Lie group with a flat left-invariant Riemannian metric is necessarily Abelian.

Since a nilpotent Lie group with rational structure constants has a (co-compact) lattice [94], we obtain.

Corollary 3.5.6 *The groups* $(\mathcal{L}(\eta), h)$ *admit lattices* $\Gamma \subset \mathcal{L}(\eta)$, *provided that* η *has rational coefficients with respect to some basis.* $M = M(\eta, \Gamma) := \Gamma \backslash \mathcal{L}(\eta)$ *is a flat compact homogeneous pseudo-Riemannian manifold. The connected component of the identity in the isometry group of* M *is the image of the natural group homomorphism* π *from* $\mathcal{L}(\eta)$ *into the isometry group of* M.

Proof First we remark that the bi-invariant metric h induces an $\mathcal{L}(\eta)$-invariant metric on the homogeneous space $M = \Gamma \backslash \mathcal{L}(\eta)$. Let G be the connected component of the identity in the isometry group of $(\mathcal{L}(\eta), h) \cong \mathbb{R}^{k,l}$. The connected component of the identity in the isometry group of M is the image of the centraliser $Z_G(\Gamma)$ of Γ in G under the natural homomorphism $Z_G(\Gamma) \to Isom(M)$. Now the statement about the isometry group follows from the fact that the centraliser of the left-action of $\Gamma \subset \mathcal{L}(\eta)$ on $\mathcal{L}(\eta)$ is precisely the right-action of $\mathcal{L}(\eta)$ on $\mathcal{L}(\eta)$, since $\Gamma \subset \mathcal{L}(\eta)$ is Zariski-dense, see Theorem 2.1 of [101] . $\square$

3.6 Classification Results for Flat Nearly ε-Kähler Manifolds

3.6.1 Classification Results for Flat Nearly Pseudo-Kähler Manifolds

In this section we denote by $\mathbb{C}^{k,l}$ the complex vector space $(\mathbb{C}^n, J_{can})$, $n = k + l$, endowed with the standard J_{can}-invariant pseudo-Euclidean scalar product g_{can} of signature $(2k, 2l)$.

Let (M, g, J) be a flat nearly pseudo-Kähler manifold. Then there exists for each point $p \in M$ an open set $U_p \subset M$ containing the point p, a connected open set U_0 of $\mathbb{C}^{k,l}$ containing the origin $0 \in \mathbb{C}^{k,l}$ and an isometry

$$\Phi : (U_p, g) \tilde{\to} (U_0, g_{can}),$$

such that at the point p we have:

$$\Phi_* J_p = J_{can} \Phi_*.$$

In other words, we can suppose, that locally M is a connected open subset of $\mathbb{C}^{k,l}$ containing the origin 0 and that $g = g_{can}$ and $J_0 = J_{can}$.

Proposition 3.6.1 *Let (M, g, J) be a flat nearly pseudo-Kähler manifold. Then*

1) $\eta_X \circ \eta_Y = 0$ for all X, Y,
2) $\nabla \eta = \bar{\nabla} \eta = 0$.

Proof From the curvature identity (3.15) we have for $X, Y, Z, W \in TM$

$$0 = R(W, X, Y, Z) - R(W, X, JY, JZ) = g((\nabla_X J)Y, (\nabla_Z J)W) = -g((\nabla_Z J)(\nabla_X J)Y, W)$$

$$= -g(J(\nabla_Z J)J(\nabla_X J)Y, W) = -4g(\eta_Z \eta_X Y, W).$$

This shows $\eta_X \circ \eta_Y = 0$ for all $X, Y \in TM$ and finishes the proof of part 1). The second part follows from 1) and $\bar{\nabla} \eta = 0$. In fact, one has

$$(\nabla_X \eta)_Y = (\bar{\nabla}_X \eta)_Y + \eta \eta_X Y + [\eta_X, \eta_Y] = 0, \text{ for } X, Y \in TM,$$

which shows part 2). $\qquad\qquad\qquad\qquad\qquad\qquad\qquad\qquad\qquad\qquad\qquad\qquad\square$

From Theorem 3.1.5 and Proposition 3.6.1 we obtain.

Corollary 3.6.2 *Let $M \subset \mathbb{C}^{k,l}$ be an open neighborhood of the origin endowed with a nearly pseudo-Kähler structure (g, J) such that $g = g_{can}$ and $J_0 = J_{can}$. Then the $(1, 2)$-tensor*

$$\eta := \frac{1}{2} J \nabla J$$

defines a constant three-form on $M \subset \mathbb{C}^{k,l} = \mathbb{R}^{2k,2l}$ defined by

$$\eta(X, Y, Z) := g(\eta_X Y, Z)$$

satisfying

(i) $\eta_X \eta_Y = 0$, $\forall X, Y \in TM$,
(ii) $\{\eta_X, J_{can}\} = 0$, $\forall X \in TM$.

Conversely, we have the next Lemma.

Lemma 3.6.3 (Lemma 5 of [41]) *Let η be a constant three-form on an open connected neighbourhood $M \subset \mathbb{C}^{k,l}$ of 0 satisfying (i) and (ii) of Corollary 3.6.2. Then there exists a unique almost complex structure J on M such that*

a) $J_0 = J_{can}$,
b) $\{\eta_X, J\} = 0$, $\forall X \in TM$,
c) $DJ = -2J\eta$,

where D stands for the Levi-Civita connection of the pseudo-Euclidian vector space $\mathbb{C}^{k,l}$. With $\bar{\nabla} := D - \eta$ and assuming b), the last equation is equivalent to

c') $\bar{\nabla}J = 0$.

More precisely, the almost complex structure is given by the formula

$$J = \exp\left(2 \sum_{i=1}^{2n} x^i \eta_{\partial_i}\right) J_{can}$$

$$\overset{(i)}{=} \left(Id + 2 \sum_{i=1}^{2n} x^i \eta_{\partial_i}\right) J_{can}, \tag{3.52}$$

where x^i are linear coordinates of $\mathbb{C}^{k,l} = \mathbb{R}^{2k,2l} = \mathbb{R}^{2n}$ and $\partial_i = \frac{\partial}{\partial x^i}$.

Theorem 3.6.4 *Let η be a constant three-form on a connected open set $U \subset \mathbb{C}^{k,l}$ containing 0 which satisfies (i) and (ii) of Corollary 3.6.2. Then there exists a unique almost complex structure given by Eq. (3.52) on U such that*

a) $J_0 = J_{can}$,
b) $M(U, \eta) := (U, g = g_{can}, J)$ is a flat nearly pseudo-Kähler manifold.

Any flat nearly pseudo-Kähler manifold is locally isomorphic to a flat nearly pseudo-Kähler manifold of the form $M(U, \eta)$.

Now we discuss the general form of solutions of (i) and (ii) of Corollary 3.6.2. In the following we shall freely identify the real vector space $V := \mathbb{C}^{k,l} = \mathbb{R}^{2k,2l} = \mathbb{R}^{2n}$ with its dual V^* by means of the pseudo-Euclidian scalar product $g = g_{can}$.

Let us recall, that the support of $\eta \in \Lambda^3 V$ is defined by

$$\Sigma_\eta := \mathrm{span}\{\eta_X Y \mid X, Y \in V\} \subset V.$$

Proposition 3.6.5 *A three-form $\eta \in \Lambda^3 V^* \cong \Lambda^3 V$ satisfies (i) of Corollary 3.6.2 if and only if there exists an isotropic subspace $L \subset V$ such that $\eta \in \Lambda^3 L \subset \Lambda^3 V$. If η satisfies (i) and (ii) of Corollary 3.6.2 then there exists a J_{can}-invariant isotropic subspace $L \subset V$ with $\eta \in \Lambda^3 L$.*

Proof The proposition follows from Lemmata 2.3.2 and 2.3.4 of Chap. 2 by taking $L = \Sigma_\eta$. □

Remark 3.6.6 From the Proposition 3.6.5 we conclude that there are no strict flat nearly pseudo-Kähler manifolds of dimension less than 8. We shall see later that the dimension cannot be smaller than 12, see Corollary 3.6.8.

In the following we set $\Lambda^- W := [\![\Lambda^{3,0} W + \Lambda^{0,3} W]\!]$, where W is a complex vector space.

Theorem 3.6.7 *A three-form $\eta \in \Lambda^3 V^* \cong \Lambda^3 V$ satisfies (i) and (ii) of Corollary 3.6.2 if and only if there exists an isotropic J_{can}-invariant subspace $L \subset V$ such that $\eta \in \Lambda^- L \subset \Lambda^3 L \subset \Lambda^3 V$. (The smallest such subspace L is Σ_η.)*

Proof By Proposition 3.6.5, the conditions (i) and (ii) of Corollary 3.6.2 imply the existence of an isotropic J_{can}-invariant subspace $L \subset V$ such that $\eta \in \Lambda^3 L$. By Lemma 2.3.3 of Chap. 2 the condition (ii) is equivalent to $\eta \in \Lambda^- V$. Therefore $\eta \in \Lambda^3 L \cap \Lambda^- V = \Lambda^- L$. The converse statement follows from the same argument. □

Corollary 3.6.8 *There are no strict flat nearly pseudo-Kähler manifolds of dimension less than 12.*

Proof By Theorems 3.6.4 and 3.6.7 any flat nearly pseudo-Kähler manifold M is locally of the form $M(U, \eta)$, where $\eta \in \Lambda^- L$ for an isotropic J_{can}-invariant subspace $L \subset V$ and $U \subset V$ is an open subset. $M(U, \eta)$ is strict if and only if $\eta \neq 0$, which is possible only for $\dim_{\mathbb{C}} L \geq 3$, i.e. for $\dim M \geq 12$. □

Theorem 3.6.9 *Any strict flat nearly pseudo-Kähler manifold is locally a pseudo-Riemannian product $M = M_0 \times M(U, \eta)$ of a flat pseudo-Kähler factor M_0 of maximal dimension and a strict flat nearly pseudo-Kähler manifold $M(U, \eta)$ of (real) signature $(2m, 2m)$, $4m = \dim M(U, \eta) \geq 12$. The J_{can}-invariant isotropic support Σ_η has complex dimension m.*

Proof By Theorems 3.6.4 and 3.6.7, M is locally isomorphic to an open subset of a manifold of the form $M(V, \eta)$, where $\eta \in \Lambda^3 V$ has a J_{can}-invariant and isotropic support $L = \Sigma_\eta$. We choose a J_{can}-invariant isotropic subspace $L' \subset V$ such that $V' := L + L'$ is nondegenerate and $L \cap L' = 0$ and put $V_0 = (L + L')^\perp$. Then $\eta \in \Lambda^3 V' \subset \Lambda^3 V$ and $M(V, \eta) = M(V_0, 0) \times M(V', \eta)$. Notice that $M(V_0, 0)$ is simply the flat pseudo-Kähler manifold V_0 and that $M(V', \eta)$ is strict and of split signature $(2m, 2m)$, where $m = \dim_{\mathbb{C}} L \geq 3$. □

Corollary 3.6.10 *Let (M, g, J) be a flat nearly Kähler manifold with a (positive or negative) definite metric g then $\eta = 0$, $\bar{\nabla} = D$ and $DJ = 0$, i.e. (M, g, J) is a Kähler manifold.*

For the rest of this section we consider the case $V \cong \mathbb{C}^{m,m}$ and denote a maximal J_{can}-invariant isotropic subspace by L. We will say that a complex three-form $\zeta \in \Lambda^3(\mathbb{C}^m)^*$ has *maximal support* if $\text{span}\{\zeta(Z, W, \cdot)|Z, W \in \mathbb{C}^m\} = (\mathbb{C}^m)^*$.

Corollary 3.6.11 *Any non-zero complex three-form $\zeta \in \Lambda^{3,0}L \cong \Lambda^3(\mathbb{C}^m)^*$ defines a complete flat simply connected strict nearly pseudo-Kähler manifold $M(\eta) := M(V, \eta)$, $\eta = \zeta + \bar{\zeta} \in \Lambda^3 L \subset \Lambda^3 V$, of split signature. $M(\eta)$ has no pseudo-Kähler de Rham factor if and only if ζ has maximal support.*

Conversely, any complete flat simply connected nearly pseudo-Kähler manifold without pseudo-Kähler de Rham factor is of this form.

Proof This follows from the previous results observing that the support of η is maximally isotropic if and only if ζ has maximal support.   □

Corollary 3.6.12 *The map $\zeta \mapsto M(\zeta + \bar{\zeta})$ induces a bijective correspondence between $GL_m(\mathbb{C})$-orbits on the open subset $\Lambda^3_{reg}(\mathbb{C}^m)^* \subset \Lambda^3(\mathbb{C}^m)^*$ of three-forms ζ with maximal support and isomorphism classes of complete flat simply connected nearly pseudo-Kähler manifolds $M(\zeta + \bar{\zeta})$ of real dimension $4m \geq 12$ and without pseudo-Kähler de Rham factor.*

Example 3.6.13

1) The case $m \leq 5$.

   For $m = 3, 4, 5$ the group $GL_m(\mathbb{C})$ acts transitively on $\Lambda^3_{reg}(\mathbb{C}^m)^* = \Lambda^3(\mathbb{C}^m)^* \setminus \{0\}$. Therefore there exists precisely one complete flat simply connected strict nearly pseudo-Kähler manifold of dimension 12, 16 and 20 respectively.

2) The case $m = 6$.

   $GL_6(\mathbb{C})$ has precisely one open orbit in $\Lambda^3_{reg}(\mathbb{C}^6)^*$. This orbit consists of the stable three-forms $\Lambda^3_{stab}(\mathbb{C}^6)^*$ in the sense of Hitchin [75], cf. Sect. 2.1 in Chap. 2. We may recall, that a three-form ζ on $\mathbb{C}^6$ is stable if and only if $\zeta = e_1^* \wedge e_2^* \wedge e_3^* + e_4^* \wedge e_5^* \wedge e_6^*$ for some basis $(e_1, e_2, \ldots, e_6)$ of $\mathbb{C}^6$. $\Lambda^3(\mathbb{C}^6)^* \setminus \Lambda^3_{stab}(\mathbb{C}^6)^*$ is precisely the zero-set of the unique homogeneous quartic $SL_6(\mathbb{C})$-invariant and we have the following strict inclusions:

$$\Lambda^3_{stab}(\mathbb{C}^6)^* \subset \Lambda^3_{reg}(\mathbb{C}^6)^* \subset \Lambda^3(\mathbb{C}^6)^* \setminus \{0\}.$$

An example of an instable regular form is

$$e_1^* \wedge e_2^* \wedge e_3^* + e_1^* \wedge e_4^* \wedge e_5^* + e_2^* \wedge e_4^* \wedge e_6^*.$$

3.6.2 Classification of Flat Nearly Para-Kähler Manifolds

In this subsection we consider (C^n, τ_{can}) endowed with the standard τ_{can}-anti-invariant pseudo-Euclidian scalar product g_{can} of signature (n, n).

Let (M, g, τ) be a flat nearly para-Kähler manifold. Then there exists for each point $p \in M$ an open set $U_p \subset M$ containing the point p, a connected open set U_0 of C^n containing the origin $0 \in C^n$ and an isometry $\Phi : (U_p, g) \tilde{\to} (U_0, g_{can})$, such that in $p \in M$ we have $\Phi_* \tau_p = \tau_{can} \Phi_*$. In other words, we can suppose, that locally M is a connected open subset of C^n containing the origin 0 and that $g = g_{can}$ and $\tau_0 = \tau_{can}$.

Proposition 3.6.14 *Let (M, g, τ) be a flat nearly para-Kähler manifold. Then*

1) $\eta_X \circ \eta_Y = 0$ for all $X, Y \in TM$,
2) $\nabla \eta = \bar{\nabla} \eta = 0$.

Summarising Theorem 3.1.5 and Proposition 3.6.14 we obtain the next Corollary.

Corollary 3.6.15 *Let $M \subset C^n$ be an open neighbourhood of the origin endowed with a nearly para-Kähler structure (g, τ) such that $g = g_{can}$ and $\tau_0 = \tau_{can}$. The $(1, 2)$-tensor*

$$\eta := -\frac{1}{2}\tau D\tau$$

defines a constant three-form on $M \subset C^n = \mathbb{R}^{n,n}$ given by $\eta(X, Y, Z) = g(\eta_X Y, Z)$ and satisfying

(i) $\eta \in \mathcal{C}(V)$, i.e. $\eta_X \eta_Y = 0$, $\quad \forall X, Y \in TM$,
(ii) $\{\eta_X, \tau_{can}\} = 0$, $\quad \forall X \in TM$.

The rest of this subsection is devoted to the local classification result. In Sect. 3.6.2 we study the structure of the subset of $\mathcal{C}(V)$ given by the condition (ii) in more detail and give global classification results. The converse statement of Corollary 3.6.15 is given in the next lemma.

Lemma 3.6.16 (Lemma 2.10 of [43]) *Let η be a constant three-form on an open connected neighbourhood $M \subset C^n$ of the origin 0 satisfying (i) and (ii) of Corollary 3.6.15. Then there exists a unique para-complex structure τ on M such that*

a) $\tau_0 = \tau_{can}$,
b) $\{\eta_X, \tau\} = 0$, $\quad \forall X \in TM$,
c) $D\tau = -2\tau\eta$,

where D is the Levi-Civita connection of the pseudo-Euclidian vector space C^n.
    Let $\bar{\nabla} := D - \eta$ and assume b) then c) is equivalent to

c)' $\bar{\nabla}\tau = 0$.

Furthermore, this para-complex structure τ is skew-symmetric with respect to g_{can}. In fact, one shows, that the para-complex structure τ is given by the following formula

$$\tau = \exp\left(2\sum_{i=1}^{2n} x^i \, \eta_{\partial_i}\right)\tau_{can} \overset{(i)}{=} \left(Id + 2\sum_{i=1}^{2n} x^i \, \eta_{\partial_i}\right)\tau_{can}, \qquad (3.53)$$

where x^i are linear coordinates of $C^n = \mathbb{R}^{n,n} = \mathbb{R}^{2n}$ and $\partial_i = \frac{\partial}{\partial x^i}$.

Theorem 3.6.17 *Let η be a constant three-form on a connected open set $U \subset C^n$ containing the origin 0 which satisfies (i) and (ii) of Corollary 3.6.15. Then there exists a unique almost para-complex structure*

$$\tau = \exp\left(2\sum_{i=1}^{2n} x^i \, \eta_{\partial_i}\right)\tau_{can} \qquad (3.54)$$

on U such that a) $\tau_0 = \tau_{can}$, and b) $M(U,\eta) := (U, g = g_{can}, \tau)$ is a flat nearly para-Kähler manifold. Any flat nearly para-Kähler manifold is locally isomorphic to a flat nearly para-Kähler manifold of the form $M(U,\eta)$.

The Variety $\mathcal{C}_\tau(V)$

Now we discuss the solution of (i) and (ii) of Corollary 3.6.15. In the following we shall freely identify the real vector space $V := C^n = \mathbb{R}^{n,n} = \mathbb{R}^{2n}$ with its dual V^* by means of the pseudo-Euclidian scalar product $g = g_{can}$. The geometric interpretation is given in terms of an affine variety $\mathcal{C}_\tau(V) \subset \Lambda^3 V$.

Proposition 3.6.18 *A three-form $\eta \in \Lambda^3 V^* \cong \Lambda^3 V$ satisfies (i) of Corollary 3.6.15, i.e. $\eta_X \circ \eta_Y = 0, X, Y \in V$, if and only if there exists an isotropic subspace $L \subset V$ such that $\eta \in \Lambda^3 L \subset \Lambda^3 V$. If η satisfies (i) and (ii) of Corollary 3.6.15 then there exists a τ_{can}-invariant isotropic subspace $L \subset V$ with $\eta \in \Lambda^3 L$.*

Proof The proposition follows from Lemmata 2.3.2 and 2.3.4 of Chap. 2 by taking $L = \Sigma_\eta$. □

A three-form η on a para-complex vector space (W, τ_{can}) decomposes with respect to the grading induced by the decomposition $W^{1,0} \oplus W^{0,1}$ into $\eta = \eta^+ + \eta^-$. In the remainder of this subsection we set for convenience $\eta^+ \in \Lambda^+ W := \Lambda^{2,1}W + \Lambda^{1,2}W$ and $\eta^- \in \Lambda^- W := \Lambda^{3,0}W + \Lambda^{0,3}W$.

Theorem 3.6.19 *A three-form $\eta \in \Lambda^3 V^* \cong \Lambda^3 V$ satisfies (i) and (ii) of Corollary 3.6.15 if and only if there exists an isotropic τ_{can}-invariant subspace L such that $\eta \in \Lambda^- L = \Lambda^{3,0}L + \Lambda^{0,3}L \subset \Lambda^3 L \subset \Lambda^3 V$ (The smallest such subspace L is Σ_η.).*

Proof By Proposition 3.6.18, the conditions (i) and (ii) of Corollary 3.6.15 imply the existence of an isotropic τ_{can}-invariant subspace $L \subset V$ such that $\eta \in \Lambda^3 L$.

By Lemma 2.3.3 of Chap. 2 the condition (ii) is equivalent to $\eta \in \Lambda^- V$. Therefore $\eta \in \Lambda^3 L \cap \Lambda^- V = \Lambda^- L$. The converse statement follows from the same argument.

$\square$

Corollary 3.6.20

(i) *The conical affine variety*

$$\mathcal{C}_\tau(V) := \{\eta \mid \eta \text{ satisfies } (i) \text{ and } (ii)\} \subset \Lambda^3 V$$

has the following description

$$\mathcal{C}_\tau(V) = \bigcup_{L \subset V} \Lambda^- L = \bigcup_{L \subset V} (\Lambda^3 L^+ + \Lambda^3 L^-),$$

where the union is over all τ-invariant maximal isotropic subspaces.

(ii) *If $\dim V < 12$ then it holds $\mathcal{C}_\tau(V) = \Lambda^3 V^+ \cup \Lambda^3 V^-$.*
(iii) *Any flat nearly para-Kähler manifold M is locally of the form $M(U, \eta)$, for some $\eta \in \mathcal{C}_\tau(V)$ and some open subset $U \subset V$.*
(iv) *There are no strict flat nearly para-Kähler manifolds of dimension less than 6.*

Proof

(i) follows from Theorem 3.6.19.
(ii) Let $L \subset V$ be a τ-invariant isotropic subspace. If $\dim V < 12$, then $\dim L < 6$ and, hence, either $\dim L^+ < 3$ or $\dim L^- < 3$. In the first case we have

$$\Lambda^- L = \Lambda^3 L^+ + \Lambda^3 L^- = \Lambda^3 L^- \subset \Lambda^3 V^-,$$

in the second case it is $\Lambda^- L = \Lambda^3 L^+ + \Lambda^3 L^- = \Lambda^3 L^+ \subset \Lambda^3 V^+$.
(iii) is a consequence of (i), Theorems 3.6.17 and 3.6.19.
(iv) By (iii) the strict flat nearly para-Kähler manifold M is locally of the form $M(U, \eta)$, which is strict if and only if $\eta \neq 0$. This is only possible for $\dim L \geq 3$, i.e. for $\dim M \geq 6$.

$\square$

Example 3.6.21 We have the following example which shows that part (ii) of Corollary 3.6.20 fails in dimension ≥ 12:

Consider $(V, \tau) = (C^6, i_\varepsilon) = \mathbb{R}^6 \oplus i_\varepsilon \mathbb{R}^6$, for $\varepsilon = 1$, with a basis given by $(e_1^+, \ldots, e_6^+, e_1^-, \ldots, e_6^-)$, such that $e_i^\pm$ form a basis of $V^\pm$ with $g(e_i^+, e_j^-) = \delta_{ij}$. Then the form $\eta := e_1^+ \wedge e_2^+ \wedge e_3^+ + e_4^- \wedge e_5^- \wedge e_6^-$ lies in the variety $\mathcal{C}_\tau(V)$.

Theorem 3.6.22 *Any strict flat nearly para-Kähler manifold is locally a pseudo-Riemannian product $M = M_0 \times M(U, \eta)$ of a flat para-Kähler factor M_0 of maximal dimension and a flat nearly para-Kähler manifold $M(U, \eta)$, $\eta \in \mathcal{C}_\tau(V)$, of signature (m, m), $2m = \dim M(U, \eta) \geq 6$ such that Σ_η has dimension m.*

Proof By Theorems 3.6.17 and 3.6.19, M is locally isomorphic to an open subset of a manifold of the form $M(V, \eta)$, where $\eta \in \Lambda^3 V$ has a τ_{can}-invariant and isotropic support $L = \Sigma_\eta$. We choose a τ_{can}-invariant isotropic subspace $L' \subset V$ such that $V' := L + L'$ is nondegenerate and $L \cap L' = 0$ and put $V_0 = (L + L')^\perp$. Then $\eta \in \Lambda^3 V' \subset \Lambda^3 V$ and $M(V, \eta) = M(V_0, 0) \times M(V', \eta)$. Notice that $M(V_0, 0)$ is simply the flat para-Kähler manifold V_0 and that $M(V', \eta)$ is strict of split signature (m, m), where $m = \dim L \geq 3$.   □

Corollary 3.6.23 *Any simply connected nearly para-Kähler manifold with a (geodesically) complete flat metric is a pseudo-Riemannian product $M = M_0 \times M(\eta)$ of a flat para-Kähler factor $M_0 = \mathbb{R}^{l,l}$ of maximal dimension and a flat nearly para-Kähler manifold $M(\eta) := M(V, \eta)$, $\eta \in C_\tau(V)$, of signature (m, m) such that Σ_η has dimension $m = 0, 3, 4, \ldots$.*

Next we wish to describe the moduli space of (complete simply connected) flat nearly para-Kähler manifolds M of dimension $2n$ up to isomorphism. Without restriction of generality we will assume that $M = M(\eta)$ has no para-Kähler de Rham factor, which means that $\eta \in C_\tau(V)$ has maximal support Σ_η, i.e. $\dim \Sigma_\eta = n$. We denote by $C_\tau^{reg}(V) \subset C_\tau(V)$ the open subset consisting of elements with maximal support. The group

$$G := \operatorname{Aut}(V, g_{can}, \tau_{can}) \cong GL(n, \mathbb{R})$$

acts on $C_\tau(V)$ and preserves $C_\tau^{reg}(V)$. Two nearly para-Kähler manifolds $M(\eta)$ and $M(\eta')$ are isomorphic if and only if η and η' are related by an element of the group G.

For $\eta \in C_\tau(V)$ we denote by p, q the dimensions of the eigenspaces of τ on Σ_η for the eigenvalues $1, -1$, respectively. We call the pair $(p, q) \in \mathbb{N}_0 \times \mathbb{N}_0$ the **type** of η. We will also say that the corresponding flat nearly para-Kähler manifold $M(\eta)$ has type (p, q). We denote by $C_\tau^{p,q}(V)$ the subset of $C_\tau(V)$ consisting of elements of type (p, q). Notice that $p + q \leq n$ with equality if and only if $\eta \in C_\tau^{reg}(V)$. We have the following decomposition

$$C_\tau^{reg}(V) = \bigcup_{(p,q) \in \Pi} C_\tau^{p,q}(V),$$

where $\Pi := \{(p, q) | p, q \in \mathbb{N}_0 \setminus \{1, 2\}, p + q = n\}$. The group $G = GL(n, \mathbb{R})$ acts on the subsets $C_\tau^{p,q}(V)$ and we are interested in the orbit space $C_\tau^{p,q}(V)/G$.

Fix a τ-invariant maximally isotropic subspace $L \subset V$ of type (p, q) and put $\Lambda_{reg}^- L := \Lambda^- L \cap C_\tau^{reg}(V) \subset C_\tau^{p,q}(V)$. The stabiliser $G_L \cong GL(L^+) \times GL(L^-) \cong GL(p, \mathbb{R}) \times GL(q, \mathbb{R})$ of $L = L^+ + L^-$ in G acts on $\Lambda_{reg}^- L$.

Theorem 3.6.24 *There is a natural one-to-one correspondence between complete simply connected flat nearly para-Kähler manifolds of type (p, q), $p + q = n$, and the points of the following orbit space:*

$$C_\tau^{p,q}(V)/G \cong \Lambda_{reg}^- L/G_L \subset \Lambda^- L/G_L = \Lambda^3 L^+/GL(L^+) \times \Lambda^3 L^-/GL(L^-).$$

Proof Consider two complete simply connected flat nearly para-Kähler manifolds M, M'. By the previous results we can assume that $M = M(\eta)$, $M' = M(\eta')$ are associated with $\eta, \eta' \in C_\tau^{p,q}(V)$. It is clear that M and M' are isomorphic if η and η' are related by an element of G. To prove the converse we assume that $\varphi : M \to M'$ is an isomorphism of nearly para-Kähler manifolds. By the results of Sect. 3.5 η defines a simply transitive group of isometries. This group preserves also the para-complex structure τ, which is ∇-parallel and hence left-invariant. This shows that M and M' admit a transitive group of automorphisms. Therefore, we can assume that φ maps the origin in $M = V$ to the origin in $M' = V$. Now φ is an isometry of pseudo-Euclidian vector spaces preserving the origin. Thus φ is an element of $O(V)$ preserving also the para-complex structure τ and hence $\varphi \in G$.

The identification of orbit spaces can be easily checked using Lemma 2.3.2 and the fact that any τ-invariant isotropic subspace $\Sigma = \Sigma^+ + \Sigma^-$ can be mapped onto L by an element of G.   □

3.7 Conical Ricci-Flat Nearly Para-Kähler Manifolds

Definition 3.7.1 A conical semi-Riemannian manifold (M, g, ξ) is a semi-Riemannian manifold (M, g) endowed with a vector field ξ such that

$$\nabla\xi = \mathrm{Id}, \tag{3.55}$$

where ∇ is the Levi-Civita connection of g. It is called regular, if the function $k := g(\xi, \xi)$ has no zeros.

A conical nearly para-Kähler manifold (M, τ, g, ξ) is a nearly para-Kähler manifold (M, τ, g) such that (M, g, ξ) is conical and a conical para-Kähler manifold (M, P, g, ξ) is a para-Kähler manifold (M, P, g) such that (M, g, ξ) is conical. For a proof of the following Proposition we refer to Proposition 6 of [39].

Proposition 3.7.2 *Let (M, g, ξ) be a regular conical semi-Riemannian manifold. Then the level sets $M_c := \{k = c\}, c \in \mathbb{R}$, are smooth hypersurfaces perpendicular to ξ or empty. If $M_c \neq \emptyset$, then g induces a semi-Riemannian metric g_c on M_c.*

Theorem 3.7.3 *Let (M, τ, g, ξ) be a Ricci-flat conical (strict) nearly para-Kähler manifold and define an endomorphism field P by*

$$P := \left(\mathrm{Id} + \frac{1}{4}N_\xi\right) \circ \tau. \tag{3.56}$$

(i) If $N(X, Y, Z)$ has isotropic support, then (M, P, g) is a para-Kähler manifold.
(ii) If the real dimension of M is 6, then (M, P, g) is a para-Kähler manifold.

We remark that the tuple (g, ξ) remains the same and in consequence (M, g, ξ) is conical and Ricci-flat. Hence (M, τ, g, ξ) is a Ricci-flat conical para-Kähler manifold.

Proof It suffices to show (i), since by Corollary 3.1.11 the Nijenhuis tensor $N(X, Y, Z)$ has isotropic support in dimension 6. Using $\{N_\xi, \tau\} = 0$, where N_ξ was defined as the endomorphism field given by $N_\xi Y = N(\xi, Y)$, we compute

$$P^2 = \left(\left(\mathrm{Id} + \frac{1}{4} N_\xi \right) \circ \tau \right)^2 = \left(\mathrm{Id} + \frac{1}{4} N_\xi \right) \circ \left(\mathrm{Id} - \frac{1}{4} N_\xi \right) \circ \tau^2$$

$$= \mathrm{Id} - \frac{1}{16} N_\xi \circ N_\xi = \mathrm{Id}.$$

Moreover, the condition

$$g(PX, Y) = -g(X, PY)$$

follows, since $N_\xi \circ \tau$ is skew-symmetric. In fact, we have

$$g(N_\xi \tau X, Y) = g(N(\xi, \tau X), Y) = 4g(\tau (\nabla_\xi \tau) \tau X, Y) = -4g((\nabla_\xi \tau) X, Y),$$

which is skew-symmetric in X, Y. Hence (M, P, g) defines an almost para-Hermitian structure. To show that it is para-Kähler we determine

$$(\nabla_X P) Y = \nabla_X \left(\left(\mathrm{Id} + \frac{1}{4} N_\xi \right) \circ \tau \right) Y = (\nabla_X \tau) Y + \frac{1}{4} \nabla_X (N_\xi \circ \tau) Y$$

$$= (\nabla_X \tau) Y + \frac{1}{4} \left[(\nabla_X N_\xi) \tau Y + N_\xi (\nabla_X \tau) Y \right]$$

$$= (\nabla_X \tau) Y + \frac{1}{4} N_{\nabla_X \xi} (\tau Y) = (\nabla_X \tau) Y - \frac{1}{4} \tau N_X Y = 0.$$

In this computation we used that $N(\cdot, \cdot, \cdot)$ has isotropic support and that $N(X, Y)$ is ∇-parallel (by Lemma 3.1.13). Namely, it is

$$(\nabla_X N_\xi) W = \nabla_X (N(\xi, W)) - N(\xi, \nabla_X W) = (\nabla_X N)(\xi, W) + N(\nabla_X \xi, W) = N_{\nabla_X \xi} W.$$

The statement that $N(X, Y, Z)$ is ∇-parallel is also shown in Lemma 3.1.13 and it does not vanish if (M, τ, g) is strict nearly para-Kähler. □

Remark 3.7.4 As the attentive reader observes, the ansatz $P = \left(\mathrm{Id} + \frac{1}{4} N_\xi \right) \circ \tau$ yields an almost para-complex structure, if N is of type $(3, 0) + (0, 3)$ and has isotropic support. This structure is para-Kähler if and only if it holds $N_{\nabla_X \xi} Y = 4\tau(\nabla_X \tau) Y$, i.e. $N_X Y = N_{\nabla_X \xi} Y$. If M is strict nearly para-Kähler, this implies $\nabla_X \xi = X$. This means ξ needs to be conical.

Theorem 3.7.5 *Let (M, P, g, ξ) be a Ricci-flat conical para-Kähler manifold of (real) dimension $2m$ carrying a (non-vanishing) parallel 3-form $\varphi(X, Y, Z)$ of type $(3, 0) + (0, 3)$ with isotropic support and define an endomorphism field τ by*

$$\tau = \left(\mathrm{Id} - \frac{1}{4}\varphi_\xi \right) \circ P. \tag{3.57}$$

Then (M, τ, g) is a (strict) nearly para-Kähler manifold.
As the pair (g, ξ) is not changed, (M, g, ξ) is conical and Ricci-flat. In consequence (M, τ, g, ξ) is a (strict) Ricci-flat conical nearly para-Kähler manifold.

Proof Here the endomorphism field φ_X is given by $\varphi_X Y := g^{-1} \circ \varphi(X, Y, \cdot)$.[9] Since $\varphi(X, Y, Z)$ has type $(3, 0) + (0, 3)$ one has $\{\varphi_\xi, P\} = 0$ (this follows from Eq. (2.35)) and we compute as before

$$\tau^2 = \left(\left(\mathrm{Id} - \frac{1}{4}\varphi_\xi \right) \circ P \right)^2 = \left(\mathrm{Id} - \frac{1}{16}\varphi_\xi \circ \varphi_\xi \right) \circ P^2 = \mathrm{Id}.$$

The last step follows, since $\varphi(X, Y, Z)$ has isotropic support (cf. the proof of Corollary 3.1.11 (b)). By the type condition it is $\varphi(\xi, PX, Y) = \varphi(P\xi, X, Y) = -\varphi(P\xi, Y, X)$ which means that $\varphi_\xi \circ P$ is skew-symmetric. From this it follows $g(\tau X, Y) = -g(X, \tau Y)$. It is left to check the nearly para-Kähler condition

$$(\nabla_X \tau)Y = \nabla_X \left(\left(\mathrm{Id} - \frac{1}{4}\varphi_\xi \right) \circ P \right) Y = (\nabla_X P)Y - \frac{1}{4}\nabla_X(\varphi_\xi \circ P)Y$$

$$= -\frac{1}{4} \left[(\nabla_X \varphi_\xi) \circ PY + \varphi_\xi (\nabla_X P)Y \right] = -\frac{1}{4}\varphi_{\nabla_X \xi}(PY) = \frac{1}{4}P(\varphi_X Y),$$

which is skew-symmetric, since $\varphi(X, Y, Z)$ is a 3-form. Hence (M, τ, g) is a nearly para-Kähler manifold. If $\varphi(X, Y, Z)$ is non-vanishing, then (M, τ, g) is strict nearly para-Kähler. $\square$

Remark 3.7.6 One may choose $\lambda\varphi(\cdot, \cdot, \cdot)$, $0 \neq \lambda \in \mathbb{R}$, in place of $\varphi(\cdot, \cdot, \cdot)$. Geometrically this corresponds to rescaling the conical vector field ξ by the factor λ.

Remark 3.7.7 Let us make an observation concerning Theorems 3.7.3 and 3.7.5. The Ansatz for τ only gives a para-complex structure, if it is $\varphi_\xi \circ \varphi_\xi = 0$. This implies, that $\varphi(X, Y, Z)$ has isotropic support and a para-Hermitian metric has automatically split signature. Therefore we only can have these examples for indefinite metrics (compare also Remark 3.7.11 for more comments).
In the following, we suppose that ξ is space-like, i.e. it is $g(\xi, \xi) > 0$. We can always achieve this by replacing the metric g by $-g$. Since $\pm g$ have the same Levi-Civita connection, this operation is compatible with the nearly para-Kähler condition (3.1)

---

[9]Here g^{-1} is the inverse of the map $g : TM \to T^*M$, $X \mapsto g(X, \cdot)$.

and the conical condition (3.55). One observes that one always can assume $M_1 = \{g(\xi, \xi) = 1\} \neq \emptyset$ after rescaling g by a positive constant without violating neither the nearly para-Kähler nor the conical condition.

Proposition 3.7.8 *Let (M, P, g, ξ) be a regular conical para-Kähler manifold with $M_1 \neq \emptyset$, then M_1 with the induced metric g_1 and the Reeb vector field $T = P\xi_{|M_1}$ is a para-Sasaki manifold.*

The manifold M is Ricci-flat if and only if M_1 is an Einstein manifold with scalar curvature $2m(2m + 1)$.

Proof The conical vector field ξ is regular and Proposition 3.7.2 implies that (M_1, g_1) is a semi-Riemannian manifold. Denote by ∇^1 the Levi-Civita connection of g_1. By construction T is time-like, i.e. $g_1(T, T) = -1$ and tangential (see Proposition 3.7.2). Moreover, T is a Killing vector field, since one has for vector fields X, Y on M_1

$$\mathcal{L}_T g_1(X, Y) = g_1(\nabla_X^1 T, Y) + g_1(X, \nabla_Y^1 T) = g(\nabla_X P\xi, Y) + g(X, \nabla_Y P\xi)$$
$$= g(P\nabla_X \xi, Y) + g(X, P\nabla_Y \xi) = g(PX, Y) + g(X, PY) = 0.$$

Additionally T is geodesic, since for $X \in TM_1$ it is

$$g_1(\nabla_T^1 T, X) = g(\nabla_T T, X) = g(\nabla_T P\xi, X) = g(PT, X) = 0.$$

Since T is a Killing vector field $\Phi := \nabla^1 T$ is skew-symmetric and we have

$$\Phi X = \nabla_X^1 T = (\nabla_X T)^{tan} = (PX)^{tan} = PX - g(\xi, PX)\xi = PX + g_1(T, X)\xi,$$

where $\cdot^{tan}$ is the projection on TM_1. This means $\Phi T = 0$ and $\Phi X = PX$ for $X \in TM_1$ perpendicular to T. It follows

$$\Phi^2(X) = P\Phi X + g_1(T, \Phi X)\xi = P\Phi X = X + g_1(T, X)T.$$

We compute $(\nabla_X^1 \Phi)Y$ for $Y = T$

$$(\nabla_X^1 \Phi)T = \nabla_X^1(\Phi T) - \Phi(\nabla_X^1 T) = -\Phi^2(X) = -X - g_1(T, X)T$$

and for Y perpendicular to T

$$(\nabla_X^1 \Phi)Y = \nabla_X^1(\Phi Y) - \Phi(\nabla_X^1 Y)$$
$$\overset{(*)}{=} P(\nabla_X^1 Y) - g(\xi, P\nabla_X^1 Y)\xi - \Phi(\nabla_X^1 Y) - g_1(X, Y)T$$
$$\overset{(**)}{=} -g_1(X, Y)T.$$

For $(*)$ we used

$$\nabla_X^1(\Phi Y) = \nabla_X^1(PY) = (\nabla_X(PY))^{tan} = (P\nabla_X Y)^{tan} = (P(\nabla_X Y)^{tan})^{tan} + g(\nabla_X Y, \xi)T$$
$$= (P(\nabla_X^1 Y))^{tan} - g(Y, \nabla_X \xi)T = P(\nabla_X^1 Y) - g(\xi, P\nabla_X^1 Y)\xi - g_1(X, Y)T$$

and $(**)$ follows from

$$P(\nabla_X^1 Y) - g(\xi, P\nabla_X^1 Y)\xi - \Phi(\nabla_X^1 Y) = g_1(P\xi, \nabla_X^1 Y)\xi - g_1(T, \nabla_X^1 Y)\xi = 0.$$

Summarizing it holds

$$(\nabla_U^1 \Phi)V = -g_1(U, V)T + g_1(V, T)U.$$

Hence we have checked all the conditions of Definition 2.6.1 and conclude that M_1 is a para-Sasaki manifold. In consequence the cone $\widehat{M}_1$ is a para-Kähler manifold, which is Einstein if and only if (M_1, g_1) is Einstein and the scalar curvature of g_1 equals $2m(2m + 1)$ (cf. Remark 2.6.2 (i) and (ii)).   □

Theorem 3.7.9 *Let (N^5, g, T) be a para-Sasaki Einstein manifold of dimension 5 and denote by $(M^6 = \widehat{N}, \widehat{g}, P, \xi)$ the associated conical Ricci-flat para-Kähler manifold on the cone $M = \widehat{N}$ over N, then the cone M can be endowed with the structure of a conical Ricci-flat strict nearly para-Kähler six-manifold $(M, \tau, \widehat{g}, \xi)$. Moreover, M is flat if and only if N has constant curvature.*

Proof By Remark 2.6.2 (ii) the cone $\widehat{N}$ is a conical Ricci-flat para-Kähler six-manifold $(\widehat{N}, \widehat{g}, P, \xi)$ and hence admits a non-vanishing parallel three-form φ with isotropic support. From Theorem 3.7.5 we obtain a strict nearly para-Kähler structure τ on $\widehat{N}$ such that $(\widehat{N}, \tau, \widehat{g}, \xi)$ is a conical Ricci-flat nearly para-Kähler six-manifold. The last statement follows from the fact that $\widehat{N}$ is flat if and only if N has constant curvature.   □

In the following we call a nearly para-Kähler manifold M, which is the space-like metric cone $M = \widehat{N}$ over some semi-Riemannian manifold N a **nearly para-Kähler cone**. Summarising we have shown the following result.

Theorem 3.7.10 *There is a one to one correspondence between Ricci-flat strict nearly para-Kähler cones with isotropic Nijenhuis tensor and space-like cones over para-Sasaki Einstein manifolds endowed with a parallel 3-form having isotropic support.*

Remarks 3.7.11

(a) An analogous Ansatz can be made in almost complex geometry.

(i) In this setting one still needs a form with isotropic support. Since non-trivial three-forms with isotropic support do not exist for Riemannian metrics, the Ansatz does only give something new, i.e. non-Kähler examples, for pseudo-Riemannian metrics and real dimension $\dim M \geq 12$, cf. Theorem 3.6.7 and Corollary 3.6.8.

(ii) Further one would get a cone over a Sasaki Einstein manifold with indefinite metric. Such manifolds can for instance be obtained as T-duals of homogeneous Sasaki manifolds of real dimension at least 11. These manifolds are only classified in dimensions ≤ 7 and the classification is possibly extended to dimension 9 and 11 using [19] (see Section 11.1.1 of [20]).

Details shall be postponed to future work.

(b) When we are not insisting on irreducible examples, one has the following construction in the almost pseudo-Hermitian world: Denote by (M^m, g_M, J_M) and (N^n, g_N, J_N) two nearly Kähler Einstein-manifolds with the same Einstein constant, then the pseudo-Riemannian product $(M \times N, g_M \oplus (-g_N), J_M \oplus J_N)$ is a nearly pseudo-Kähler manifold with vanishing Ricci curvature.

3.8 Evolution of Hypo Structures to Nearly Pseudo-Kähler Six-Manifolds

3.8.1 Linear Algebra of Five-Dimensional Reductions of $\mathrm{SU}(1, 2)$-Structures

In this short section we prepare the linear algebra of dimensional reductions.

Lemma 3.8.1 *Let V be a six-dimensional real vector space and $(\omega, \rho) \in \Lambda^2 V^* \times \Lambda^3 V^*$ a compatible normalised pair of stable forms. Denote by $h = h_{(\omega,\rho)}$ the induced metric, let $N \in V$ be a unit vector with $h(N, N) = -\varepsilon \in \{\pm 1\}$ and denote by $W = N^\perp$ the orthogonal complement of $\mathbb{R} \cdot N$. Then the quadruple $(\eta, \omega_1, \omega_2, \omega_3)$ defined by*

$$\eta = \beta (N \lrcorner \omega), \quad \omega_1 = \alpha \, \omega|_W, \quad \omega_2 = N \lrcorner J_\rho^* \rho, \quad \omega_3 = -N \lrcorner \rho \tag{3.58}$$

with $\alpha, \beta \in \{\pm 1\}$ defines an $\mathrm{SU}^\varepsilon(p, q)$-structure with $p + q = 2$ on W. Moreover, one has

$$\omega = \alpha \, \omega_1 + \beta \, n \wedge \eta,$$

$$\rho = -\varepsilon\beta \, \eta \wedge \omega_2 - n \wedge \omega_3,$$

$$J_\rho^* \rho = -\varepsilon\beta \, \eta \wedge \omega_3 + n \wedge \omega_2,$$

where $n \in V^*$ is the dual of N and

$$\eta \wedge \omega_2 = -\varepsilon\beta \, \rho|_W \text{ and } \eta \wedge \omega_3 = -\varepsilon\beta \, J_\rho^* \rho|_W.$$

Proof As above (see Eqs. (2.7) and (2.20) of Chap. 2) we may choose a basis $\{e_1, \ldots, e_6\}$ of V, such that the stable forms ω, ρ and $J_\rho^* \rho$ are given in the normal forms

$$\omega = -e^{12} - e^{34} + e^{56},$$
$$\rho = e^{135} - (e^{146} + e^{236} + e^{245})$$

with $\lambda(\rho) = -4\nu^{\otimes 2}$ for $\nu = e^{123456} > 0$. Furthermore, it holds $J_\rho e_i = -e_{i+1}$, $J_\rho e_{i+1} = e_i$ for $i \in \{1, 3, 5\}$ and

$$J_\rho^* \rho = e^{246} - (e^{235} + e^{145} + e^{136}).$$

1) In the case $\varepsilon = 1$ we can suppose $N = e_1$ and obtain

$$\eta = -\beta \, e^2, \quad \alpha \, \omega_1 = -e^{34} + e^{56}, \quad \omega_2 = -e^{36} - e^{45}, \quad \omega_3 = -e^{35} + e^{46}.$$

One easily sees

$$\omega_1^2 = -\omega_2^2 = -\omega_3^2 = -2e^{3456} \text{ and } \eta \wedge \omega_j^2 = 2\beta \, e^{23456} \neq 0$$

and $\omega_j \wedge \omega_k = 0$ for $1 \leq j < k \leq 3$.

Moreover, one gets

$$\omega = \alpha \, \omega_1 + \beta \, n \wedge \eta,$$
$$\rho = -\beta \, \eta \wedge \omega_2 - n \wedge \omega_3 \text{ and}$$
$$J_\rho^* \rho = -\beta \, \eta \wedge \omega_3 + n \wedge \omega_2,$$

where n is the dual of N. Further, one has

$$\eta \wedge \omega_2 = -\beta \, \rho_{|W} \text{ and } \eta \wedge \omega_3 = -\beta \, J_\rho^* \rho_{|W}.$$

2) For $\varepsilon = -1$ we can choose $N = e_5$ and get

$$\eta = \beta \, e^6, \quad \alpha \, \omega_1 = -e^{12} - e^{34}, \quad \omega_2 = -e^{14} - e^{23}, \quad \omega_3 = -e^{13} + e^{24}.$$

One easily sees

$$\omega_1^2 = \omega_2^2 = \omega_3^2 = 2e^{1234} \text{ and } \eta \wedge \omega_j^2 = 2\beta \, e^{12346} \neq 0$$

and $\omega_j \wedge \omega_k = 0$ for $1 \leq j < k \leq 3$.

Moreover, one gets

$$\omega = \alpha \, \omega_1 + \beta \, n \wedge \eta,$$

$$\rho = \beta \, \eta \wedge \omega_2 - n \wedge \omega_3 \quad \text{and}$$

$$J_\rho^* \rho = \beta \, \eta \wedge \omega_3 + n \wedge \omega_2,$$

where n is the dual of N. Finally, one has

$$\eta \wedge \omega_2 = \beta \, \rho_{|W} \quad \text{and} \quad \eta \wedge \omega_3 = \beta \, J_\rho^* \rho_{|W}.$$

$\square$

3.8.2   Evolution of Hypo Structures

A five-manifold N^5 carries an $\mathrm{SU}^\varepsilon(p,q)$-structure with $p+q = 2$ provided, that its frame bundle admits a reduction to $\mathrm{SU}^\varepsilon(p,q)$. For the group $\mathrm{SU}(2)$ it is shown in [36], that such a structure is determined by a quadruple of differential forms $(\omega_1, \omega_2, \omega_3, \eta)$. We shortly derive the analogous statement for our setting. Let $f \colon N^5 \to M^6$ be an oriented hypersurface in a six-manifold M^6 endowed with an $\mathrm{SU}(1,2)$-structure given by a triple (ω, ψ^+, ψ^-) of compatible stable forms (cf. Sect. 2.1).

This $\mathrm{SU}(1,2)$-structure induces an $\mathrm{SU}^\varepsilon(p,q)$-structure with $p+q = 2$ on N^5 via the definitions

$$\omega_1 = \alpha f^* \omega, \quad b^- \omega_2 = v \lrcorner \psi^-, \quad b^+ \omega_3 = v \lrcorner \psi^+, \quad \eta = \beta \, v \lrcorner \omega, \tag{3.59}$$

where $\alpha, \beta, b^+, b^- \in \{\pm 1\}$ are real constants and v denotes the unit normal vector field of N^5 of length $\varepsilon = -g(v,v)$.

In case, that the holonomy of M^6 is contained in $SU(1,2)$ or in other words the $SU(1,2)$-structure is integrable, which is equivalent to the equations

$$d\omega = 0, \quad d\psi^+ = 0 \quad \text{and} \quad d\psi^- = 0 \tag{3.60}$$

we obtain a hypo structure on N in the sense of the next Definition.

Definition 3.8.2 An $\mathrm{SU}^\varepsilon(p,q)$-structure with $p+q = 2$ determined by $(\eta, \omega_1, \omega_2, \omega_3)$ is called hypo provided, that it satisfies

$$d\omega_1 = 0, \quad d(\eta \wedge \omega_2) = 0 \quad \text{and} \quad d(\eta \wedge \omega_3) = 0.$$

For the Riemannian case the next lemma is shown in [36].

Lemma 3.8.3 *Let* $f : N^5 \to M^6$ *be an oriented hypersurface in a six-manifold* M^6 *endowed with an integrable* $\mathrm{SU}(1, 2)$-*structure, then the induced* $\mathrm{SU}^\varepsilon(p, q)$-*structure, given by (3.59) and with* $p + q = 2$, *on* N^5 *is a hypo-structure.*

Proof From $\alpha f^* \omega = \omega_1$ one has $d\omega_1 = \alpha\, d(f^* \omega) = \alpha f^*(d\omega) = 0$. Moreover, with help of the Lemma 3.8.1 one has

$$b^+ f^* \psi^+ = -\varepsilon\beta\, \eta \wedge \omega_2 \text{ and } b^- f^* \psi^- = -\varepsilon\beta\, \eta \wedge \omega_3,$$

which implies using (3.60)

$$-\varepsilon\beta\, d(\eta \wedge \omega_3) = b^-\, d(f^* \psi^-) = b^- f^* d\psi^- = 0$$
$$= b^+\, d(f^* \psi^+) = b^+ f^* d\psi^+ = -\varepsilon\beta\, d(\eta \wedge \omega_2).$$

This shows, that the induced $\mathrm{SU}^\varepsilon(p, q)$-structure is a hypo structure on N^5. □

Starting with an $\mathrm{SU}^\varepsilon(p, q)$-structure with $p + q = 2$ on N^5 determined by $(\eta, \omega_1, \omega_2, \omega_3)$ we define a two-form

$$\omega = \alpha\omega_1 + \varepsilon\beta\, dt \wedge \eta \tag{3.61}$$

and three-forms $\psi^\pm$ on $N^5 \times \mathbb{R}$ by

$$\psi^+ = a^+ \eta \wedge \omega_2 + b^+ dt \wedge \omega_3, \quad \psi^- = a^- \eta \wedge \omega_3 + b^- dt \wedge \omega_2, \tag{3.62}$$

where t is the coordinate on $\mathbb{R}$ and $\alpha, \beta, a^\pm, b^\pm \in \{\pm 1\}$ are non-zero real constants. Note, that the $a^\pm$ are determined from α, β and $b^\pm$ by Lemma 3.8.1. Then a partial converse of the result of the last Lemma is given in the next Proposition.

Proposition 3.8.4 *One can lift a hypo* $\mathrm{SU}^\varepsilon(p, q)$-*structure with* $p + q = 2$ *to an integrable* $\mathrm{SU}(1, 2)$-*structure on* $N \times \mathbb{R}$ *if it belongs to a one-parameter family of* $\mathrm{SU}^\varepsilon(p, q)$-*structures* $(\eta(t), \omega_1(t), \omega_2(t), \omega_3(t))$, *where* t *is the coordinate on* $\mathbb{R}$, *satisfying the Conti-Salamon type evolution equations*

$$\partial_t \omega_1 = \varepsilon\beta\alpha\, d^5\eta, \tag{3.63}$$

$$\partial_t(\eta \wedge \omega_3) = a^- b^-\, d\omega_2, \tag{3.64}$$

$$\partial_t(\eta \wedge \omega_2) = a^+ b^+\, d\omega_3. \tag{3.65}$$

Proof In this proof we write d^5 and d^6 for the exterior differentials on N^5 and $M^6 = N \times \mathbb{R}$. With (3.61) it follows, that

$$0 = d^6\omega = \alpha d^5\omega_1 + (\alpha\partial_t\omega_1 - \varepsilon\beta\, d^5\eta) \wedge dt.$$

This is equivalent to $d^5\omega_1 = 0$ and $\partial_t\omega_1 = \varepsilon\beta\alpha\, d^5\eta$, i.e. Eq. (3.63). From the definition of ψ^+ one gets, that

$$0 = d^6\psi^+ = a^+ d^5(\omega_2 \wedge \eta) + dt \wedge \left(a^+\partial_t(\omega_2 \wedge \eta) - b^+ d^5\omega_3\right)$$

is equivalent to $d^5(\omega_2 \wedge \eta) = 0$ and $\partial_t(\omega_2 \wedge \eta) = a^+ b^+ d^5\omega_3$, i.e. Eq. (3.65). For ψ^- as given in (3.62) we obtain, that

$$0 = d^6\psi^- = a^- d^5(\omega_3 \wedge \eta) + dt \wedge (a^-\partial_t(\omega_3 \wedge \eta) - b^- d^5\omega_2)$$

itself is equivalent to $d^5(\omega_3 \wedge \eta) = 0$ and Eq. (3.64). □

Examples of this type are given by the pseudo-Riemannian cousins of Sasaki-Einstein manifolds, namely para-Sasaki-Einstein and Lorentzian-Sasaki-Einstein manifolds. These can be characterised by the fact, that the space-like/time-like cone is a Ricci-flat Kähler-Einstein manifold or equivalently this cone has an integrable $SU(1, 2)$-structure. Here one considers the special solution of the above evolution equations on $N^5 \times \mathbb{R}$ given by

$$\omega = t^2\alpha\omega_1 + t\varepsilon\beta dt \wedge \eta, \tag{3.66}$$

$$\psi^+ = a^+ t^3\eta \wedge \omega_2 + t^2 b^+ dt \wedge \omega_3, \tag{3.67}$$

$$\psi^- = a^- t^3\eta \wedge \omega_3 + t^2 b^- dt \wedge \omega_2. \tag{3.68}$$

The integrability conditions read

$$0 = d\omega = d(t^2\alpha\omega_1 + t\varepsilon\beta dt \wedge \eta) = tdt \wedge (2\alpha\omega_1 - \varepsilon\beta d\eta),$$

$$0 = d\psi^+ = d(a^+ t^3\eta \wedge \omega_2 + t^2 b^+ dt \wedge \omega_3)$$

$$= t^2 dt \wedge (3a^+\eta \wedge \omega_2 - b^+ d^5\omega_3) + t^3 a^+ d^5(\eta \wedge \omega_2),$$

$$0 = d\psi^- = d(a^- t^3\eta \wedge \omega_3 + t^2 b^- dt \wedge \omega_2)$$

$$= t^2 dt \wedge (3a^-\eta \wedge \omega_3 - b^- d^5\omega_2) + t^3 a^- d^5(\eta \wedge \omega_3).$$

This is equivalent to the para-Sasaki-Einstein or Lorentz-Sasaki-Einstein equations

$$d\eta = 2\varepsilon\alpha\,\beta\,\omega_1, \quad d\omega_2 = 3b^- a^-\omega_3 \wedge \eta, \quad d\omega_3 = 3b^+ a^+\omega_2 \wedge \eta. \tag{3.69}$$

Obviously, Eq. (3.69) imply the hypo equations.

The next result has been discovered in the Riemannian case in [53].

Proposition 3.8.5 *Let $f: N^5 \to M^6$ be a totally geodesic oriented hypersurface in a nearly pseudo-Kähler manifold M^6 with unit normal vector field v, then the induced $SU^\varepsilon(p, q)$-structure with $p + q = 2$ and $\varepsilon = -g(v, v)$ satisfies the hypo equations.*

Proof From the first nearly Kähler equation one has

$$d\omega_1 = \alpha d(f^*\omega) = \alpha f^* d\omega = 3\alpha f^* \psi^+ = 3(\alpha a^+)\eta \wedge \omega_2. \qquad (3.70)$$

Moreover, we compute

$$\beta^{-1} d\eta = d(v \lrcorner \omega) = (\mathcal{L}_v \omega) - v \lrcorner d\omega = (\mathcal{L}_v \omega) - 3v \lrcorner \psi^+ = (\mathcal{L}_v \omega) - 3b^+ \omega_3,$$

where $\mathcal{L}_v$ is the Lie-derivative. Since N^5 is totally geodesic, it is $(\mathcal{L}_v \omega) = \nabla_v \omega$. Using $\bar{\nabla}\omega = 0$ we get with $\bar{\nabla} = \nabla + \frac{1}{2}T$

$$\nabla_v \omega(X, Y) = \bar{\nabla}_v \omega(X, Y) + \frac{1}{2}[\omega(T(v, X), Y) + \omega(X, T(v, Y))]$$

$$= \omega(T(v, X), Y) = b^+ \omega_3(X, Y),$$

as with $\omega(\cdot, \cdot) = g(\cdot, J\cdot)$ it is

$$\omega(T(v, X), Y) = -g(J(\nabla_v J)X, JY) = g(X, (\nabla_v J)Y) = \psi^+(v, X, Y) = b^+ \omega_3(X, Y),$$

which shows

$$d\eta = -2\beta b^+ \omega_3.$$

Finally, we compute with help of $d\psi^- = -2\omega \wedge \omega$, i.e. the second nearly Kähler equation

$$b^- d\omega_2 = d(v \lrcorner \psi^-) = \mathcal{L}_v \psi^- - v \lrcorner d\psi^- = \nabla_v \psi^- + 2v \lrcorner (\omega \wedge \omega)$$

$$= \bar{\nabla}_v \psi^- - \frac{1}{4} v \lrcorner \sum_k \sigma_k(e_k \lrcorner \psi^-) \wedge (e_k \lrcorner \psi^-) + 2v \lrcorner (\omega \wedge \omega)$$

$$\overset{(*)}{=} -\frac{1}{4} v \lrcorner (2\omega \wedge \omega) + 2v \lrcorner (\omega \wedge \omega) = \frac{3}{2} v \lrcorner (\omega \wedge \omega) = 3(\alpha\beta)\eta \wedge \omega_1,$$

where $\{e_1, \dots, e_6 = v\}$ is some adapted basis and where in $(*)$ we used

$$\sum_{k=1}^{6} \sigma_k(e_k \lrcorner \psi^-) \wedge (e_k \lrcorner \psi^-) = 2\omega \wedge \omega,$$

which holds for a nearly pseudo-Kähler six-manifold. $\square$

Theorem 3.8.6 *Any totally geodesic hypersurface N^5 of a nearly pseudo-Kähler six-manifold M^6 carries a hypo structure given by the quadruplet $(\tilde{\eta}, \tilde{\omega}_1, \tilde{\omega}_2, \tilde{\omega}_3) = (\eta, -\varepsilon\omega_3, \omega_2, \omega_1)$, and in consequence the Conti-Salamon type evolution equations can be solved on $N \times \mathbb{R}$.*

Proof By the last Lemma we obtain a hypo $SU^\varepsilon(p,q)$-structure $(\tilde{\eta}, \tilde{\omega}_1, \tilde{\omega}_2, \tilde{\omega}_3)$, which is a solution of the first and the third para-Sasaki-Einstein or Lorentz-Sasaki-Einstein equations (3.69). Using $b^+ = -1$ and $a^+ = -\beta$ (by Lemma 3.8.1) and setting $\alpha = -1$ yields

$$d\tilde{\eta} = d\eta = -2(\beta b^+)\omega_3 = 2\beta\,\omega_3 \overset{!!}{=} 2\,\varepsilon(\beta\alpha)\,\tilde{\omega}_1,$$

$$d\tilde{\omega}_3 = d\omega_1 \overset{(3.70)}{=} 3(\alpha a^+)\,\eta \wedge \omega_2 = -3(\alpha\beta)\,\tilde{\eta}\wedge\tilde{\omega}_2 \overset{!!}{=} 3(b^+ a^+)\tilde{\eta}\wedge\tilde{\omega}_2,$$

since again by Lemma 3.8.1,[10] it is $b^+ a^+ = \beta = -a^- b^-$. The remaining para-Sasaki-Einstein or Lorentz-Sasaki-Einstein equation

$$d\tilde{\omega}_2 = d\omega_2 = 3(b^-\alpha\beta)\eta\wedge\omega_1 = 3(\alpha\beta)\tilde{\eta}\wedge\tilde{\omega}_3 \overset{!!}{=} 3(b^-a^-)\tilde{\eta}\wedge\tilde{\omega}_3$$

holds true using $b^- = 1$ (by Lemma 3.8.1). $\qquad\qquad\qquad\qquad\qquad\square$

3.8.3 Evolution of Nearly Hypo Structures

In this subsection we generalise results of [53] to construct examples of nearly pseudo-Kähler manifolds via the nearly hypo evolution equations.

Definition 3.8.7 An $SU^\varepsilon(p,q)$-structure with $p + q = 2$ determined by $(\eta, \omega_1, \omega_2, \omega_3)$ is called **nearly hypo** provided, that it satisfies the conditions

$$d\omega_1 = 3\alpha a^+\eta\wedge\omega_2, \qquad d(\eta\wedge\omega_3) = -2a^-\omega_1\wedge\omega_1. \qquad (3.71)$$

Proposition 3.8.8 *An $SU^\varepsilon(p,q)$-structure $(\eta, \omega_1, \omega_2, \omega_3)$ with $p + q = 2$ can be lifted to a nearly pseudo-Kähler structure $(\omega(t), \psi^+(t), \psi^-(t))$ on $N^5 \times \mathbb{R}$ defined in (3.61) and (3.62) if and only if it is a nearly hypo structure which generates a 1-parameter family of $SU^\varepsilon(p,q)$-structures $(\eta(t), \omega_k(t))$ satisfying the following nearly hypo evolution equations*

$$\partial_t\omega_1 = 3b^+\alpha\,\omega_3 + \varepsilon\beta\alpha\,d\eta,$$

$$\partial_t(\eta\wedge\omega_3) = a^-b^-\,d\omega_2 - 4\varepsilon a^-\alpha\beta\omega_1\wedge\eta, \qquad (3.72)$$

$$\partial_t(\eta\wedge\omega_2) = a^+b^+\,d\omega_3.$$

---

[10]Observe, that there is a relative factor ε between dual and the metric dual of v.

Remark 3.8.9 Setting $\beta = 1$ and with $b^- = 1 = -b^+$ and $b^+ a^+ = -b^- a^- = \beta = 1$ (by Lemma 3.8.1) we get

$$\partial_t(\alpha\omega_1) = -3\,\omega_3 + \varepsilon\,d\eta,$$
$$\partial_t(\eta \wedge \omega_3) = -d\omega_2 + 4\varepsilon\eta \wedge (\alpha\omega_1), \qquad (3.73)$$
$$\partial_t(\eta \wedge \omega_2) = d\omega_3.$$

For $(\tilde\eta, \tilde\omega_1, \tilde\omega_2, \tilde\omega_3) = (-\eta, \alpha\omega_1, \omega_2, \omega_3)$, this yields

$$\partial_t\tilde\omega_1 = -3\,\tilde\omega_3 - \varepsilon\,d\tilde\eta,$$
$$\partial_t(\tilde\eta \wedge \tilde\omega_3) = d\tilde\omega_2 - 4\varepsilon\tilde\eta \wedge \tilde\omega_1, \qquad (3.74)$$
$$\partial_t(\tilde\eta \wedge \tilde\omega_2) = -d\tilde\omega_3.$$

Hence the exterior differential system on the modified data $(\tilde\eta, \tilde\omega_1, \tilde\omega_2, \tilde\omega_3)$ looks formally the same as the system found in the Riemannian case [53].

Proof From the definitions of ω_1, ω_2 and η we get, that the first nearly Kähler equation is equivalent to

$$d\omega = d(\alpha\omega_1 + \varepsilon\beta dt \wedge \eta) = \alpha d\omega_1 + dt \wedge (\alpha\partial_t\omega_1 - \varepsilon\beta d\eta)$$
$$= 3\psi^+ = 3a^+\eta \wedge \omega_2 + 3b^+ dt \wedge \omega_3,$$

i.e. $d\omega_1 = 3\alpha a^+\eta \wedge \omega_2$ and $\partial_t\omega_1 = 3b^+\alpha\,\omega_3 + \varepsilon\beta\alpha\,d\eta$. For the second nearly Kähler equation we have

$$d\psi^- = d(a^-\eta \wedge \omega_3 + b^- dt \wedge \omega_2) = a^- d(\eta \wedge \omega_3) + dt \wedge (a^-\partial_t(\eta \wedge \omega_3) - b^- d\omega_2)$$
$$= -2(\alpha\omega_1 + \varepsilon\beta dt \wedge \eta)^2 = -2\,\omega_1 \wedge \omega_1 - 4\varepsilon\alpha\beta\,dt \wedge \omega_1 \wedge \eta,$$

which is equivalent to

$$d(\eta \wedge \omega_3) = -2a^-\omega_1 \wedge \omega_1 \quad \text{and} \quad \partial_t(\eta \wedge \omega_3) = a^- b^- d\omega_2 - 4\varepsilon a^-\alpha\beta\omega_1 \wedge \eta.$$

These are the first two evolution equations and the nearly hypo equations. The third equation is needed to show, that the nearly hypo property is conserved along the evolution. Firstly, one has

$$\partial_t(d\omega_1 - 3\alpha a^+\eta \wedge \omega_2) = d(\partial_t\omega_1) - 3\alpha a^+\partial_t(\eta \wedge \omega_2)$$
$$= d(3b^+\alpha\,\omega_3 + \varepsilon\beta\alpha\,d\eta) - 3\alpha a^+\partial_t(\eta \wedge \omega_2)$$
$$= 3b^+\alpha\,d\omega_3 - 3\alpha a^+\partial_t(\eta \wedge \omega_2),$$

which vanishes by the third evolution equation $\partial_t(\eta \wedge \omega_2) = a^+ b^+ d\omega_3$. For the other nearly hypo equation we compute

$$
\begin{aligned}
\partial_t[d(\eta \wedge \omega_3) + 2a^- \omega_1 \wedge \omega_1] &= d[\partial_t(\eta \wedge \omega_3)] + 2a^- \partial_t(\omega_1 \wedge \omega_1) \\
&= d[a^- b^- d\omega_2 - 4\varepsilon a^- \alpha\beta \omega_1 \wedge \eta] \\
&\quad + 4a^- \partial_t(\omega_1) \wedge \omega_1 \\
&= -4\varepsilon a^- \alpha\beta d(\omega_1 \wedge \eta) \\
&\quad + 4a^- \omega_1 \wedge (3b^+ \alpha\, \omega_3 + \varepsilon\beta\alpha\, d\eta) \\
&= -4\varepsilon a^- \alpha\beta(d(\omega_1 \wedge \eta) - \omega_1 \wedge d\eta) \\
&= -4\varepsilon a^- \alpha\beta\, d\omega_1 \wedge \eta = 0,
\end{aligned}
$$

where we used, that by the already proven first hypo equation $d\omega_1$ is (along the flow) a multiple of $\eta \wedge \omega_2$. Hence the nearly hypo condition is preserved along a solution of the system (3.72). $\square$

Proposition 3.8.10 *Any* $SU^\varepsilon(p, q)$-*structure with* $p + q = 2$ *satisfying the para- or pseudo-Sasaki equations* (3.69) *defines a nearly hypo structure* $(\tilde\eta, \tilde\omega_1, \tilde\omega_2, \tilde\omega_3) = (\eta, \omega_3, \omega_2, \omega_1)$.

Proof From (3.69) one has after setting $\alpha = b^+ = -1$ (by Lemma 3.8.1)

$$
d\tilde\omega_1 = d\omega_3 = 3b^+ a^+\, \omega_2 \wedge \eta = 3\alpha a^+ \tilde\eta \wedge \tilde\omega_2
$$

and

$$
d(\tilde\eta \wedge \tilde\omega_3) = d(\eta \wedge \omega_1) = d\eta \wedge \omega_1 = 2\varepsilon\alpha\beta\, \omega_1 \wedge \omega_1 = -2\alpha\beta\, \omega_3 \wedge \omega_3 \stackrel{!!}{=} -2a^- \tilde\omega_1 \wedge \tilde\omega_1,
$$

where we used $\omega_1 \wedge \omega_1 = -\varepsilon\, \omega_3 \wedge \omega_3$. This yields the claim, since one has $a^- = -\beta = \alpha\beta$ (by Lemma 3.8.1). $\square$

Proposition 3.8.11 *Let* $f: N^5 \to M^6$ *be an immersion of an oriented 5-manifold into a 6-dimensional nearly pseudo-Kähler manifold, then the induced* $SU^\varepsilon(p, q)$-*structure* $(\tilde\eta, \tilde\omega_1, \tilde\omega_2, \tilde\omega_3) = (\eta, \omega_1, \omega_2, \omega_3)$ *with* $p+q = 2$ *is a nearly hypo structure.*

Proof Let us first observe, that one has

$$
a^+ \eta \wedge \omega_2 = f^* \psi^+ \quad \text{and} \quad a^- \eta \wedge \omega_3 = f^* \psi^-,
$$

which implies

$$
\begin{aligned}
d\tilde\omega_1 = d\omega_1 &= \alpha\, d(f^*\omega) = \alpha f^*(d\omega) \\
&= 3\alpha f^* \psi^+ = 3\alpha\, a^+ \eta \wedge \omega_2 = 3\alpha a^+ \tilde\eta \wedge \tilde\omega_2,
\end{aligned}
$$

$$a^- d(\tilde{\eta} \wedge \tilde{\omega}_3) = a^- d(\eta \wedge \omega_3) = f^* d\psi^-$$
$$= -2f^* (\omega \wedge \omega) = -2\,\omega_1 \wedge \omega_1 = -2\,\tilde{\omega}_1 \wedge \tilde{\omega}_1.$$

This proves the nearly hypo equations.   □

Theorem 3.8.12 *Let* $(N^5, \eta, \omega_1^0, \omega_2^0, \omega_3^0)$ *be an* ε*-Sasaki-Einstein* $SU^\varepsilon(p,q)$*-structure, then*

$$\omega_1 = f_\varepsilon^2 \left(f_\varepsilon \omega_1^0 + f_\varepsilon' \omega_3^0 \right), \tag{3.75}$$

$$\omega_2 = f_\varepsilon^2 \omega_2^0, \tag{3.76}$$

$$\omega_3 = -f_\varepsilon^2 \left(f_\varepsilon' \omega_1^0 + \varepsilon f_\varepsilon \omega_3^0 \right), \tag{3.77}$$

$$\eta = f_\varepsilon \eta^0 \tag{3.78}$$

with

$$f_\varepsilon(t) := \sin_\varepsilon(t) = \begin{cases} \sinh(t), & \text{for } t \in \mathbb{R} \text{ and } \varepsilon = 1, \\ \sin(t), & \text{for } t \in [0, \pi] \text{ and } \varepsilon = -1 \end{cases} \quad \text{and } I_\varepsilon := \begin{cases} \mathbb{R}^* \text{ for } \varepsilon = 1, \\ (0, \pi) \text{ for } \varepsilon = -1 \end{cases}$$

is a solution of the nearly hypo evolution equations and yields a nearly pseudo-Kähler structure on $N \times I_\varepsilon$ *Metrik angeben, anpassen with metric* $g = dt^2 + f_\varepsilon^2 g^N$ *with conical singularities in* $\{0\}$ *and* $\{0, \pi\}$ *respectively.*

Proof Recall, that we have to solve (where we omit the $\tilde{\cdot}$) the following system

$$\partial_t \omega_1 = -3\,\omega_3 - \varepsilon\,d\eta,$$
$$\partial_t (\eta \wedge \omega_3) = d\omega_2 - 4\varepsilon\eta \wedge \omega_1, \tag{3.79}$$
$$\partial_t (\eta \wedge \omega_2) = -d\omega_3.$$

As Ansatz we consider the following family of $SU^\varepsilon(p,q)$-structures with $p + q = 2$

$$\omega_1 = f^2 \left(f\omega_1^0 \pm f'\omega_3^0 \right),$$
$$\omega_2 = \sigma f^2 \omega_2^0,$$
$$\omega_3 = -f^2 \left(f'\omega_1^0 \pm \varepsilon f\omega_3^0 \right),$$
$$\eta = \sigma f \eta^0,$$

where we set $f(t) = \sin_\varepsilon(t)$, which yields $f'' = \varepsilon f$ and $f'^2 - \varepsilon f^2 = 1$. First we compute $\alpha\beta = 1$

$$\partial_t \omega_1 = 3f^2 f' \omega_1^0 \pm \left(2ff'^2 + f^2 f''\right)\omega_3^0$$

$$\overset{(*)}{=} 3\left(f^2 f' \omega_1^0 \pm \varepsilon f^3 \omega_3^0\right) \pm 2f\omega_3^0$$

$$= -3\,\omega_3 \pm \varepsilon f d\eta^0 = -3\,\omega_3 \pm \sigma\varepsilon d\eta \overset{!!}{=} -3\,\omega_3 - \varepsilon d\eta,$$

where in $(*)$ we used

$$2f(f')^2 + \varepsilon f^3 = f(2f'^2 + \varepsilon f^2) = 2f(\varepsilon f^2 + 1) + \varepsilon f^3 = 3\,\varepsilon f^3 + 2f$$

and $d\eta^0 = 2\varepsilon\omega_3^0$, since N^5 is an ε-Sasaki manifold. Hence we need to fix $\sigma = \mp 1$.
Next we calculate

$$\partial_t\,(\eta \wedge \omega_2) = \partial_t\left(f^3\eta^0 \wedge \omega_2^0\right) = 3f^2 f'\,\eta^0 \wedge \omega_2^0$$

and using $d\omega_3^0 = 0$ and $d\omega_1^0 = 3a^+ b^+ \eta^0 \wedge \omega_2^0$ yields

$$-b^+ a^+ d\omega_3 = b^+ a^+\,d^5\left(f^2\left(f'\omega_1^0 \pm \varepsilon f\omega_3^0\right)\right) = b^+ a^+ f^2 f' d^5\omega_1^0 = 3f^2 f'\,\eta^0 \wedge \omega_2^0,$$

which shows $\partial_t\,(\eta \wedge \omega_2) = -b^+ a^+ d\omega_3 = -d\omega_3$. It remains to determine the evolution of $\eta \wedge \omega_3$

$$-\partial_t\,(\eta \wedge \omega_3) = \sigma\partial_t\left(\pm\varepsilon f^4\eta^0 \wedge \omega_3^0 + f^3 f'\eta^0 \wedge \omega_1^0\right)$$

$$= \sigma\left(\pm 4\varepsilon f^3 f'\eta^0 \wedge \omega_3^0 + \left(3f^2(f')^2 + f^3 f''\right)\eta^0 \wedge \omega_1^0\right)$$

$$\overset{(*)}{=} 4\varepsilon\sigma\left(\pm f^3 f'\eta^0 \wedge \omega_3^0 + f^4\eta^0 \wedge \omega_1^0\right) + 3\sigma f^2 \eta^0 \wedge \omega_1^0$$

$$= 4\varepsilon\,\eta \wedge \omega_1 + \sigma b^- a^- f^2 d\omega_2^0 \overset{b^- a^- = -1}{=} 4\varepsilon\,\eta \wedge \omega_1 - \sigma f^2 d\omega_2^0$$

$$= 4\varepsilon\,\eta \wedge \omega_1 - d\omega_2,$$

where in $(*)$ we used

$$3f^2 f'^2 + \varepsilon f^4 = 3f^2(\varepsilon f^2 + 1) + \varepsilon f^4 = 4\,\varepsilon f^4 + 3f^2.$$

This yields

$$\partial_t\,(\eta \wedge \omega_3) = -4\varepsilon\,\eta \wedge \omega_1 + d\omega_2$$

and finishes the proof of the Theorem. □

3.9 Results in the Homogeneous Case

3.9.1 Consequences for Automorphism Groups

An automorphism of an $SU^\varepsilon(p, q)$-structure on a six-manifold M is an automorphism of principal fibre bundles or equivalently, a diffeomorphism of M preserving all tensors defining the $SU^\varepsilon(p, q)$-structure. By our discussion on stable forms in Sect. 2.1 of Chap. 2, an $SU^\varepsilon(p, q)$-structure is characterised by a pair of compatible stable forms $(\omega, \rho) \in \Omega^2 M \times \Omega^3 M$. Since the construction of the remaining tensors J, ψ^- and g is invariant, a diffeomorphism preserving the two stable forms is already an automorphism of the $SU^\varepsilon(p, q)$-structure and in particular an isometry.

This easy observation has the following consequences when combined with the exterior systems of the previous section and the naturality of the exterior derivative.

Proposition 3.9.1 *Let* (ω, ψ^+) *be an* $SU^\varepsilon(p, q)$*-structure on a six-manifold* M.

(i) *If the exterior differential equation*

$$d\omega = \mu \, \psi^+$$

is satisfied for a constant $\mu \neq 0$, *then a diffeomorphism* Φ *of* M *preserving* ω *is an automorphism of the* $SU^\varepsilon(p, q)$*-structure and in particular an isometry.*

(ii) *If the exterior differential equation*

$$d\psi^- = \nu \, \omega \wedge \omega$$

is satisfied for a constant $\nu \neq 0$, *then a diffeomorphism* Φ *of* M *preserving*

(a) *the real volume form and* ψ^+,
(b) *or the real volume form and* ψ^-,
(c) *or the* ε*-complex volume form* $\Psi = \psi^+ + i_\varepsilon \psi^-$,

is an automorphism of the $SU^\varepsilon(p, q)$*-structure and in particular an isometry.*

We like to emphasise that both parts of the Proposition apply to strict nearly ε-Kähler structures of non-zero type. The same holds true for the following Proposition.

Proposition 3.9.2 *Let* $(M^6, g, J^\varepsilon, \omega)$ *be an almost* ε*-Hermitian six-manifold with totally skew-symmetric Nijenhuis tensor and* Φ *be a diffeomorphism of* M *preserving the almost* ε*-complex structure* J^ε. *Suppose, that the structure* J^ε *is quasi-integrable,*

(i) *then* Φ *is a conformal map.*
(ii) *and additionally, assume, that one has* $d\omega^2 = 0$, *then* Φ *is a homothety on connected components of* M. *If moreover,* Φ *preserves the volume, then it is an isometry.*

Proof As Φ preserves the ε-complex structure, it also preserves the Nijenhuis tensor. From Corollary 3.1.15 it follows $g(X, Y) = f \operatorname{tr}(N_X \circ N_Y)$ for some function f on M. This yields in $p \in M$

$$g_{\Phi(p)}(\Phi_* X, \Phi_* Y) = f(\Phi(p)) \operatorname{tr}(N_{\Phi_* X} \circ N_{\Phi_* Y})$$
$$= f(\Phi(p)) \operatorname{tr}\left((\Phi^* N)_X \circ (\Phi^* N)_Y\right) = f(\Phi(p)) f(p) g_p(X, Y),$$

i.e. the conformal factor is $c := f \cdot \Phi^* f$. Further let us assume, that one has $d\omega^2 = 0$. From above we know $\Phi^*(\omega \wedge \omega) = c^2(\omega \wedge \omega)$, which yields

$$0 = d(c^2 \omega \wedge \omega)) = d(c^2) \wedge \omega^2 + c^2 d\omega^2 = d(c^2) \wedge \omega^2.$$

Using, that the map $\eta \in \Lambda^1 T^* M^6 \to \eta \wedge \omega^2 \in \Lambda^5 T^* M^6$ is an isomorphism, we obtain $d(c^2) = 0$ and hence the function c is constant on connected components of M. Recall, that the metric volume is a multiple of ω^3. This implies, that one has $c = 1$. $\qquad\square$

Corollary 3.9.3 *Let $(M, J^\varepsilon, g, \omega)$ be a nearly ε-Kähler six-manifold with $\|\nabla J^\varepsilon\|^2 \neq 0$, then a diffeomorphism Φ of M preserving J^ε is an automorphism of the $\mathrm{SU}^\varepsilon(p, q)$-structure and in particular an isometry.*

Proof The second nearly ε-Kähler equation implies $d\omega^2 = 0$. Hence we obtain from Proposition 3.9.2, that one has $\Phi^*(\omega) = c\omega$, for some constant c (on each connected component) and by the first nearly pseudo-Kähler equation

$$\Phi^*(\psi^+) = \frac{d(\Phi^* \omega)}{3} = \frac{c}{3} d\omega = c\psi^+.$$

As J^ε is preserved, this yields $\Phi^*(\psi^-) = c\psi^-$ and another time using the second nearly pseudo-Kähler equation

$$c d\psi^- = \Phi^*(d\psi^-) = v\Phi^*(\omega^2) = v c^2 \omega^2,$$

forces $c = 1$. $\qquad\square$

Conversely, it is known for *complete* Riemannian nearly Kähler manifolds, that orientation-preserving isometries are automorphism of the almost Hermitian structure except for the round sphere S^6, see for instance [26, Proposition 4.1]. However, this is not true if the metric is incomplete. In [53, Theorem 3.6], a nearly Kähler structure is constructed on the incomplete sine-cone over a Sasaki-Einstein five-manifold $(N^5, \eta, \omega_1, \omega_2, \omega_3)$. In fact, the Reeb vector field dual to the one-form η is a Killing vector field which does not preserve ω_2 and ω_3. Thus, by the formulae given in [53], its lift to the nearly Kähler six-manifold is a Killing field for the sine-cone metric which does neither preserve Ψ nor ω nor J.

3.9.2   Left-Invariant Nearly ε-Kähler Structures on $\mathrm{SL}(2,\mathbb{R}) \times \mathrm{SL}(2,\mathbb{R})$

The following lemma is the key to proving the forthcoming structure result, since it considerably reduces the number of algebraic equations on the nearly ε-Kähler candidates.

Lemma 3.9.4 (Lemma 4.1 of [110]) *Denote by $(\mathbb{R}^{1,2}, \langle \cdot, \cdot \rangle)$ the vector space $\mathbb{R}^3$ endowed with its standard Minkowskian scalar-product and denote by $\mathrm{SO}_0(1,2)$ the connected component of the identity of its group of isometries. Consider the action of $\mathrm{SO}_0(1,2) \times \mathrm{SO}_0(1,2)$ on the space of real 3×3 matrices $\mathrm{Mat}(3,\mathbb{R})$ given by*

$$\Phi : \mathrm{SO}_0(1,2) \times \mathrm{Mat}(3,\mathbb{R}) \times \mathrm{SO}_0(1,2) \to \mathrm{Mat}(3,\mathbb{R})$$

$$(A, C, B) \mapsto A^t C B.$$

Then any invertible element $C \in \mathrm{Mat}(3,\mathbb{R})$ lies in the orbit of an element of the form

$$\begin{pmatrix} \alpha & x & y \\ 0 & \beta & z \\ 0 & 0 & \gamma \end{pmatrix} \quad or \quad \begin{pmatrix} 0 & \beta & z \\ \alpha & x & y \\ 0 & 0 & \gamma \end{pmatrix}$$

with $\alpha, \beta, \gamma, x, y, z \in \mathbb{R}$ and $\alpha\beta\gamma \neq 0$.

Finally, we prove our main result of this subsection which is the following theorem. By a homothety, we define the rescaling of the metric by a real number which we do not demand to be positive since we are working with all possible signatures.

Theorem 3.9.5 *Let G be a Lie group with Lie algebra $\mathfrak{sl}(2,\mathbb{R})$. Up to homothety, there is a unique left-invariant nearly ε-Kähler structure with $\|\nabla J^\varepsilon\|^2 \neq 0$ on $G \times G$. This is the nearly pseudo-Kähler structure of signature (4,2) constructed as 3-symmetric space in Sect. 3.9.4. In particular, there is no left-invariant nearly para-Kähler structure.*

Remark 3.9.6 The proof also shows that there is a left-invariant nearly ε-Kähler structure of non-zero type on $G \times H$ with $Lie(G) = Lie(H) = \mathfrak{sl}(2,\mathbb{R})$ if $G \neq H$ which is unique up to homothety *and* exchanging the orientation.

Proof More precisely, we will prove uniqueness up to equivalence of left-invariant almost ε-Hermitian structures and homothety. We will consider the algebraic exterior system

$$d\omega = 3\psi^+, \tag{3.80}$$

$$d\psi^- = 2\,\omega \wedge \omega \tag{3.81}$$

on the Lie algebra $\mathfrak{sl}(2, \mathbb{R}) \oplus \mathfrak{sl}(2, \mathbb{R})$. By Theorem 3.1.19, solutions of this system are in one-to-one correspondence to left-invariant nearly ε-Kähler structure on $G \times G$ with $\|\nabla J^\varepsilon\|^2 = 4$. This normalisation can always be achieved by applying a homothety. Furthermore, two solutions which are isomorphic under an inner Lie algebra automorphism from

$$\mathrm{Inn}(\mathfrak{sl}(2, \mathbb{R}) \oplus \mathfrak{sl}(2, \mathbb{R})) = SO_0(1, 2) \times SO_0(1, 2)$$

are equivalent under the corresponding Lie group isomorphism. Since both factors are equal, we can also lift the outer Lie algebra automorphism exchanging the two summands to the group level. In summary, it suffices to show the existence of a solution of the algebraic exterior system (3.80), (3.81) on the Lie algebra which is unique up to inner Lie algebra automorphisms and exchanging the summands.

A further significant simplification is the observation that all tensors defining a nearly ε-Kähler structure of non-zero type can be constructed out of the fundamental two-form ω with the help of the first nearly Kähler equation (3.80) and the stable form formalism described in Sect. 2.4 of Chap. 2. We break the main part of the proof into three lemmas, step by step simplifying ω under Lie algebra automorphisms in a fixed Lie bracket.

We call $\{e_1, e_2, e_3\}$ a *standard basis* of $\mathfrak{so}(1, 2)$ if the Lie bracket satisfies

$$de^1 = -e^{23}, \quad de^2 = e^{31}, \quad de^3 = e^{12}.$$

In this basis, an inner automorphism in $SO_0(1, 2)$ acts by usual matrix multiplication on $\mathfrak{so}(1, 2)$.

Lemma 3.9.7 *Let $\mathfrak{g} = \mathfrak{h} = \mathfrak{so}(1, 2)$ and let ω be a non-degenerate two-form in*

$$\Lambda^2(\mathfrak{g} \oplus \mathfrak{h})^* = \Lambda^2 \mathfrak{g}^* \oplus (\mathfrak{g} \otimes \mathfrak{h}) \oplus \Lambda^2 \mathfrak{h}^*.$$

Then we have

$$d\omega^2 = 0 \quad \Leftrightarrow \quad \omega \in \mathfrak{g} \otimes \mathfrak{h}. \tag{3.82}$$

Proof By inspecting the standard basis, we observe that all two-forms on $\mathfrak{so}(1, 2)$ are closed whereas no non-trivial 1-form is closed. Thus, when separately taking the exterior derivative of the components of ω^2 in $\Lambda^4 = (\Lambda^3 \mathfrak{g}^* \otimes \mathfrak{h}^*) \oplus (\Lambda^2 \mathfrak{g}^* \otimes \Lambda^2 \mathfrak{h}^*) \oplus (\mathfrak{g}^* \otimes \Lambda^3 \mathfrak{h}^*)$, the equivalence is easily deduced. $\square$

Lemma 3.9.8 *Let $\mathfrak{g} = \mathfrak{h} = \mathfrak{so}(1, 2)$ and let $\{e^1, e^2, e^3\}$ be a basis of $\mathfrak{g}^*$ and $\{e^4, e^5, e^6\}$ a basis of $\mathfrak{h}^*$ such that the Lie brackets are given by*

$$de^1 = -e^{23}, \quad de^2 = e^{31}, \quad de^3 = \tau e^{12} \quad and \quad de^4 = -e^{56}, \quad de^5 = e^{64}, \quad de^6 = e^{45} \tag{3.83}$$

for some $\tau \in \{\pm 1\}$. Then, every non-degenerate two-form ω on $\mathfrak{g} \oplus \mathfrak{h}$ satisfying $d\omega^2 = 0$ can be written

$$\omega = \alpha \, e^{14} + \beta \, e^{25} + \gamma \, e^{36} + x \, e^{15} + y \, e^{16} + z \, e^{26} \qquad (3.84)$$

for $\alpha, \beta, \gamma \in \mathbb{R} - \{0\}$ and $x, y, z \in \mathbb{R}$ modulo an automorphism in $SO_0(1, 2) \times SO_0(1, 2)$.

Proof We choose standard bases $\{e^1, e^2, e^3\}$ for $\mathfrak{g}$ and $\{e^4, e^5, e^6\}$ for $\mathfrak{h}$. Using the previous lemma and the assumption $d\omega^2 = 0$, we may write $\omega = \sum_{i,j=1}^3 c_{ij} e^{i(j+3)}$ for an invertible matrix $C = (c_{ij}) \in \text{Mat}(3, \mathbb{R})$. When a pair $(A, B) \in SO_0(1, 2) \times SO_0(1, 2)$ acts on the two-form ω, the matrix C is transformed to $A^t CB$. Applying Lemma 3.9.4, we can achieve by an inner automorphism that C is in one of the normal forms given in that lemma. However, an exchange of the base vectors e_1 and e_2 corresponds exactly to exchanging the first and the second row of C. Therefore, we can always write ω in the claimed normal form by adding the sign τ in the Lie bracket of the first summand $\mathfrak{g}$. $\qquad \square$

Lemma 3.9.9 (Lemma 4.6 of [110]) *Let $\{e^1, \ldots, e^6\}$ be a basis of $\mathfrak{so}(1, 2) \times \mathfrak{so}(1, 2)$ such that*

$$de^1 = -e^{23}, \quad de^2 = e^{31}, \quad de^3 = e^{12} \quad \text{and} \quad de^4 = -e^{56}, \quad de^5 = e^{64}, \quad de^6 = e^{45}. \qquad (3.85)$$

Then the only $SU^\varepsilon(p, q)$-structure (ω, ψ^+) modulo inner automorphisms and modulo exchanging the summands, which solves the two nearly ε-Kähler equations (3.80) and (3.81), is determined by

$$\omega = \frac{\sqrt{3}}{18}(e^{14} + e^{25} + e^{36}). \qquad (3.86)$$

In fact, the uniqueness, existence and non-existence statements claimed in the theorem follow directly from this lemma and formula obtained for the quartic invariant which implies $\lambda\left(\frac{1}{3}d\omega\right) < 0$.

As explained in Sect. 3.9.4, we know that there is a left-invariant nearly pseudo-Kähler structure of indefinite signature on all the groups in question. After applying a homothety, we can achieve $\|\nabla J^\varepsilon\|^2 = 4$ and this structure has to coincide with the unique structure we just constructed. Therefore, the indefinite metric has to be of signature (4,2) by our sign conventions.

We summarise the data of the unique nearly pseudo-Kähler structure in the basis (3.85) and can easily double-check the signature of the metric explicitly:

$$\omega = \frac{1}{18}\sqrt{3} \, (e^{14} + e^{25} + e^{36})$$

$$\psi^+ = \frac{1}{54}\sqrt{3} \, (e^{126} - e^{135} + e^{156} - e^{234} + e^{246} - e^{345})$$

$$\psi^- = -\frac{1}{54}(2\,e^{123} + e^{126} - e^{135} - e^{156} - e^{234} - e^{246} + e^{345} + 2\,e^{456})$$

$$J(e_1) = -\frac{1}{3}\sqrt{3}\,e_1 - \frac{2}{3}\sqrt{3}\,e_4, \quad J(e_4) = \frac{2}{3}\sqrt{3}\,e_1 + \frac{1}{3}\sqrt{3}\,e_4$$

$$J(e_2) = -\frac{1}{3}\sqrt{3}\,e_2 + \frac{2}{3}\sqrt{3}\,e_5, \quad J(e_5) = -\frac{2}{3}\sqrt{3}\,e_2 + \frac{1}{3}\sqrt{3}\,e_5$$

$$J(e_3) = -\frac{1}{3}\sqrt{3}\,e_3 + \frac{2}{3}\sqrt{3}\,e_6, \quad J(e_6) = -\frac{2}{3}\sqrt{3}\,e_3 + \frac{1}{3}\sqrt{3}\,e_6$$

$$g = \frac{1}{9}\left((e^1)^2 - (e^2)^2 - (e^3)^2 + (e^4)^2 - (e^5)^2\right.$$
$$\left. - (e^6)^2 - e^1 \cdot e^4 - e^2 \cdot e^5 - e^3 \cdot e^6\right).$$

$\square$

Observing that in [25] very similar arguments have been applied to the Lie group $S^3 \times S^3$, we find the following non-existence result.

Proposition 3.9.10 *On the Lie groups $G \times H$ with $Lie(G) = Lie(H) = \mathfrak{so}(3)$, there is neither a left-invariant nearly para-Kähler structure of non-zero type nor a left-invariant nearly pseudo-Kähler structure with an indefinite metric.*

Proof The unicity of the left-invariant nearly Kähler structure $S^3 \times S^3$ is proved in [25, Section 3], with a strategy analogous to the proof of Theorem 3.9.5. In the following, we will refer to the English version [26]. There, it is shown in the proof of Proposition 2.5, that for any solution of the exterior system

$$d\omega = 3\psi^+$$
$$d\psi^+ = -2\mu\omega^2$$

there is a basis of the Lie algebra of $S^3 \times S^3$ and a real constant α such that

$$de^1 = e^{23}, \quad de^2 = e^{31}, \quad de^3 = e^{12} \quad \text{and} \quad de^4 = e^{56}, \quad de^5 = e^{64}, \quad de^6 = e^{45},$$
$$\omega = \alpha(e^{14} + e^{25} + e^{36}).$$

In this basis, a direct computation or formula (18) in [26] show that the quartic invariant that we denote by λ is

$$\lambda = -\frac{1}{27}\alpha^4$$

with respect to the volume form e^{123456}. Therefore, a nearly para-Kähler structure cannot exist on all the Lie groups with the same Lie algebra as $S^3 \times S^3$ by Theorem 3.1.19. A nearly pseudo-Kähler structure with an indefinite metric cannot exist either, since the induced metric is always definite as computed in the second part of Lemma 2.3 in [26]. $\square$

3.9.3   Real Reducible Holonomy

Nearly pseudo-Kähler manifolds admitting a J-invariant and $\bar{\nabla}$-parallel decomposition of the tangent bundle TM are related to twistor spaces [108] (cf. [16, 98] for Riemannian metrics) and Sects. 3.3 and 3.4 of this chapter. The next Proposition considers the complementary situation, i.e. the case where TM decomposes into two sub-bundles and J interchanges these sub-bundles and generalises a result of [99] to pseudo-Riemannian metrics.

Proposition 3.9.11 *Let (M, g, J) be a complete, strict, simply connected nearly pseudo-Kähler manifold. Suppose, that TM admits an orthogonal, $\bar{\nabla}$-parallel decomposition $TM = \mathcal{V} \oplus \mathcal{V}'$ with $\mathcal{V}' = J\mathcal{V}$, then (M, g) is a homogeneous space.*

Proof For a vector field $X = JV_1$ in $\mathcal{V}' = J\mathcal{V}$, a vector field $Y = JV_2$ in TM and vector fields V_3, V_4 in $\mathcal{V}$ by the same argument as in Lemma 3.4.1 it is

$$\bar{R}(JV_1, JV_2, V_3, V_4) = g\left([\nabla_{V_3}J, \nabla_{V_4}J]JV_1, JV_2\right) - g\left((\nabla_{JV_1}J)JV_2, (\nabla_{V_3}J)V_4\right)$$

$$= g\left([\nabla_{V_3}J, \nabla_{V_4}J]V_1, V_2\right) + g\left((\nabla_{V_1}J)V_2, (\nabla_{V_3}J)V_4\right) \quad (3.87)$$

By the symmetries (3.24) and (3.25) of the curvature tensor $\bar{R}$ the last equation determines $\bar{R}$. By Proposition 3.1.7 the torsion T and ∇J are $\bar{\nabla}$-parallel. In particular, we have

$$\bar{\nabla}_U((\nabla_X J)Y) = (\nabla_X J)\bar{\nabla}_U Y + (\nabla_{\bar{\nabla}_U X} J)Y.$$

Deriving (3.87) this implies $\bar{\nabla}\bar{R} = 0$. Hence $\bar{\nabla}$ is an Ambrose-Singer connection and as M is simply connected and complete it follows, that (M, J, g) is a homogeneous space (see [116]).   □

3.9.4   3-Symmetric Spaces

The idea of a three-symmetric space is to replace the symmetry of order two as in the case of a symmetric space by a symmetry of order three. Nearly Kähler geometry on such spaces was first studied in [66, 69].

Like symmetric spaces three-symmetric spaces have a homogenous model, which we shortly resume: Let G be a connected Lie group and s an automorphism of order 3 and let $G_0^s \subset H \subset G^s$ be a subgroup contained in the fix-point set G^s of s. The differential s_* decomposes

$$\mathfrak{g} \otimes \mathbb{C} = \mathfrak{h} \otimes \mathbb{C} \oplus \mathfrak{m}^+ \oplus \mathfrak{m}^-$$

into the eigenspaces of s_* with eigenvalues 1 and $\frac{1}{2}(-1 \pm \sqrt{-3})$. With the definition $\mathfrak{m} := (\mathfrak{m}^+ \oplus \mathfrak{m}^-) \cap \mathfrak{g}$ the decomposition

$$\mathfrak{g} = \mathfrak{h} \oplus \mathfrak{m} \tag{3.88}$$

is reductive and G/H is a reductive homogenous space. The characteristic complex structure is then defined by

$$s_{*|\mathfrak{m}} = -\frac{1}{2}Id + \frac{\sqrt{3}}{2}J. \tag{3.89}$$

The choice of an $Ad(H)$-invariant and s_*-invariant (pseudo-)Euclidean scalar-product B on $\mathfrak{m}$ endows G/H with the structure of a (pseudo-)Riemannian three-symmetric space, such that B is almost Hermitian with respect to J. The next Theorem locally relates homogeneous spaces to three-symmetric spaces.

Theorem 3.9.12 *An almost pseudo-Hermitian manifold (M, J, g) is a locally three-symmetric space if and only if it is a quasi-Kähler manifold and the torsion T and the curvature tensor $\bar{R}$ of the characteristic Hermitian connection $\bar{\nabla}$ are parallel, i.e. $\bar{\nabla}T = 0$ and $\bar{\nabla}\bar{R} = 0$.*

The proof of Theorem 3.9.12 is based on the following description of locally three-symmetric spaces given in [66].

Theorem 3.9.13 *Let (M, J, g) be an almost pseudo-Hermitian manifold. Then there exists a family of local cubic diffeomorphisms $(s_x)_{x \in M}$ such that J is the induced complex structure and such that M is a three-symmetric space if and only if*

(i) *M is quasi-Kähler, i.e. one has $(\nabla_X J)Y + (\nabla_{JX} J)JY = 0$,*
(ii) *$\sigma = s_*$ preserves $\nabla^2 J$,*
(iii) *for $X, Y, Z, T \in \Gamma(TM)$ one has*

$$R(X, Y, Z, T) = R(JX, JY, Z, T) + R(JX, Y, JZ, T) \tag{3.90}$$
$$+ R(JX, Y, Z, JT),$$

(iv) *for $X, Y, Z, T \in \Gamma(TM)$ one has*

$$(\nabla_W R)(X, Y, Z, T) + (\nabla_W R)(JX, JY, JZ, JT) = 0.$$

Proof of Theorem 3.9.12 We claim that the conditions (i)-(iv) of Theorem 3.9.13 are equivalent to the following system of equations:

$$\eta_X Y + \eta_{JX} JY = 0, \tag{3.91}$$
$$\bar{\nabla}\eta = 0, \tag{3.92}$$
$$\bar{\nabla}\bar{R} = 0, \tag{3.93}$$

Let us recall the definition $\eta := \frac{1}{2}J(\nabla J)$ which yields that the condition (3.91) is equivalent to Theorem 3.9.13 part (i). From $\nabla_X J = -2J\eta_X$ it follows

$$(\nabla^2_{X,Y}J)Z = 2J(2\eta_X\eta_Y Z - (\nabla_X\eta)_Y Z) = 2J\left(2\eta_X\eta_Y Z - (\bar{\nabla}_X\eta)_Y Z - \eta_{\eta_X Y}Z + [\eta_Y, \eta_X]Z\right).$$

By Eq. (3.92) this expression only depends on η and J, which are both preserved by σ.

Conversely, we suppose that σ preserves ∇J and $\nabla^2 J$. From Proposition 3.3 of Gray [66] one obtains

$$\bar{\nabla}\eta(X, Y, Z, T) = \bar{\nabla}\eta(JX, JY, Z, T) + \bar{\nabla}\eta(JX, Y, JZ, T)$$
$$+ \bar{\nabla}\eta(JX, Y, Z, JT) = 3\bar{\nabla}\eta(JX, JY, Z, T),$$

where the last equality follows from $\bar{\nabla}J = 0$ and Eq. (3.91). This implies replacing X by JX and Y by JY

$$3\bar{\nabla}\eta(JX, JY, Z, T) = \bar{\nabla}\eta(X, Y, Z, T) = \frac{1}{3}\bar{\nabla}\eta(JX, JY, Z, T)$$

and finally we obtain $\bar{\nabla}\eta = 0$.

Moreover, the condition (3.92) implies Theorem 3.9.13 part (iii). In fact, we claim that Theorem 3.9.13 part (iii) can be re-written as $R \in \mathcal{L}_2$, where $\mathcal{L}_2$ is one of the irreducible components of the space of curvature tensors considered as a $GL(n, \mathbb{C})$ representation [52]. More precisely, in our case we identify $\mathfrak{u}(p, q)$ with $[\lambda^{1,1}]$ (instead of $\mathfrak{u}(n)$) and $\mathfrak{u}(p, q)^\perp$ (rather than $\mathfrak{u}(n)^\perp$) with $[\lambda^{2,0}]$ and then apply the results of [52] to obtain an analogous decomposition. In particular, it follows in the quasi-Kähler case, that the complement of $\mathcal{L}_2$ only depends on $\bar{\nabla}\eta$ and we conclude that Eq. (3.92) implies Theorem 3.9.13 part (iii).

It remains to relate Theorem 3.9.13 part (iv) and Eq. (3.93). From $\nabla = \bar{\nabla} + \eta$ it follows

$$\bar{R}(X, Y, Z, W) = R(X, Y, Z, W) + g([\eta_X, \eta_Y]Z, W) - g(\eta_{\eta_X Y - \eta_Y X}Z, W),$$

where we use the condition (3.92). Using a second time the condition (3.92) and $\bar{\nabla}g = 0$ we get

$$\bar{\nabla}\bar{R} = \bar{\nabla}R = \nabla R - \eta \cdot R \tag{3.94}$$

with

$$\eta \cdot R = R(\eta\cdot, \cdot, \cdot, \cdot) + R(\cdot, \eta\cdot, \cdot, \cdot) + R(\cdot, \cdot, \eta\cdot, \cdot) + R(\cdot, \cdot, \cdot, \eta\cdot).$$

Now Eq. (3.90) implies $R(JX, JY, JZ, JW) = R(X, Y, Z, W)$, see for instance Corollary 3.4 of [66]. As η and J anti-commute, it follows from (3.90)

$$(\eta \cdot R)(JX, JY, JZ, JW) = -(\eta \cdot R)(X, Y, Z, W) \tag{3.95}$$

and as $\bar{\nabla}$ is Hermitian we have

$$\bar{\nabla}R(X, Y, Z, W) = \bar{\nabla}R(JX, JY, JZ, JW). \tag{3.96}$$

Equation (3.94) and (3.96) yield

$$2(\bar{\nabla}_W \bar{R})(X, Y, Z, T) \quad = \quad (\bar{\nabla}_W \bar{R})(X, Y, Z, T) + (\bar{\nabla}_W \bar{R})(JX, JY, JZ, JT)$$
$$\overset{Eqs.\,(3.94),(3.95)}{=} (\nabla_W R)(X, Y, Z, T) + (\nabla_W R)(JX, JY, JZ, JT),$$

which shows the equivalence of Theorem 3.9.13 part (iv) and Eq. (3.93). $\square$
The following proposition relates the information coming from the Hermitian structure to the data of the homogeneous space.

One may suppose G to be simply connected, since otherwise one considers its universal cover $\pi : \tilde{G} \to G$ and the isomorphic homogeneous space $\tilde{G}/\tilde{H}$ with $\tilde{H} = \pi^{-1}(H)$.

Proposition 3.9.14 *Let $(M = G/H, J, g)$ be a (simply connected) reductive homogeneous almost pseudo-Hermitian manifold, then $M = G/H$ is three-symmetric if and only if it is quasi-Kähler and the connection $\bar{\nabla}$ coincides with the normal connection ∇^{nor} of the reductive homogeneous space G/H.*

Proof Let $(M = G/H, J, g)$ be a reductive homogeneous space with adapted reductive decomposition $\mathfrak{g} = \mathfrak{h} \oplus \mathfrak{m}$. The invariant almost complex structure J induces a complex structure on $\mathfrak{m}$ and an invariant decomposition

$$\mathfrak{m}^{\mathbb{C}} = \mathfrak{m}^{1,0} \oplus \mathfrak{m}^{0,1}. \tag{3.97}$$

The invariance of $\mathfrak{m}$ and J implies

$$[\mathfrak{h}, \mathfrak{m}^{1,0}] \subset \mathfrak{m}^{1,0} \text{ and } [\mathfrak{h}, \mathfrak{m}^{0,1}] \subset \mathfrak{m}^{0,1}. \tag{3.98}$$

The three-symmetry s is now obtained by the integration of the map σ

$$\sigma_{|\mathfrak{h}} = Id_{|\mathfrak{h}}, \ \sigma_{|\mathfrak{m}^{1,0}} = jId_{|\mathfrak{m}^{1,0}}, \sigma_{|\mathfrak{m}^{0,1}} = j^2 Id_{|\mathfrak{m}^{0,1}},$$

where $j = -\frac{1}{2}Id + \frac{\sqrt{3}}{2}i$. The map σ integrates (since G is supposed to be simply connected) to s if and only if it is an automorphism of the Lie algebra $\mathfrak{g}$. By Eq. (3.98) and the definition of σ this is the case if and only if one has

$$[\mathfrak{m}^{1,0}, \mathfrak{m}^{1,0}] \subset \mathfrak{m}^{0,1}, [\mathfrak{m}^{0,1}, \mathfrak{m}^{0,1}] \subset \mathfrak{m}^{1,0} \text{ and } [\mathfrak{m}^{1,0}, \mathfrak{m}^{0,1}] \subset \mathfrak{h}. \tag{3.99}$$

Recall, that the torsion of the normal connection (see [87, Chapter X]) is given by the invariant tensor

$$T^{nor}(u, v) = -[u, v]^{\mathfrak{m}}.$$

In terms of the torsion T^{nor} the integrability conditions (3.99) are

$$T^{nor}(u, v) = -[u, v]^{\mathfrak{m}^{0,1}}, \quad \text{for } u, v \in \mathfrak{m}^{1,0},$$

$$T^{nor}(u, v) = -[u, v]^{\mathfrak{m}^{1,0}}, \quad \text{for } u, v \in \mathfrak{m}^{0,1},$$

$$T^{nor}(u, v) = \quad 0, \quad \quad \text{for } u \in \mathfrak{m}^{1,0} \text{ and } v \in \mathfrak{m}^{0,1}.$$

In other words T^{nor} is a multiple of the Nijenhuis tensor. Contraction with the metric yields a tensor $g(T^{nor}(\cdot, \cdot), \cdot)$ which is of type $\otimes^3 \left(\mathfrak{m}^{1,0}\right)^* \oplus \otimes^3 \left(\mathfrak{m}^{0,1}\right)^*$ w.r.t. the complex structure induced by $\mathfrak{m}^{1,0} \oplus \mathfrak{m}^{0,1}$ and skew-symmetric in the first two entries. These symmetries exclude contributions of the pseudo-Riemannian version of class $\mathcal{W}_3$ and $\mathcal{W}_4$ in the Gray-Hervella list [68] and hence (M, J, g) is of type $\mathcal{W}_1 \oplus \mathcal{W}_2$, i.e. (M, J, g) is a quasi-Kähler manifold. Moreover, we have $T^{nor} \in [\![\lambda^{2,0} \otimes \lambda^{1,0}]\!]$ and we obtain $\nabla^{nor} - \nabla^g \in \mathcal{W}_1 \oplus \mathcal{W}_2 \subset T^*M \otimes \mathfrak{u}_{p,q}^\perp$. This means that ∇^{nor} is the intrinsic connection which equals the characteristic Hermitian connection. Summarizing (3.97) is the decomposition into eigenspaces of an automorphism of order three if and only if $(M = G/H, J, g)$ is quasi-Kähler and the normal connection coincides with the intrinsic connection.   □

As a consequence the torsion and the curvature of the connection $\bar{\nabla}$ are given by

$$T(u, v) = -[u, v]^{\mathfrak{m}} \text{ and } \bar{R}(u, v) = [u, v]^{\mathfrak{h}} \text{ with } u, v \in \mathfrak{m}. \tag{3.100}$$

reductive, if it holds

$$B([X, Y]^{\mathfrak{m}}, Z) = B(X, [Y, Z]^{\mathfrak{m}}) \text{ for } X, Y, Z \in \mathfrak{m}.$$

The next result was already shown in [66] Proposition 5.6 for pseudo-Riemannian metrics. It is a consequence of $\nabla^{nor} = \bar{\nabla}$ for three-symmetric spaces.

Proposition 3.9.15 *A three-symmetric space is a nearly pseudo-Kähler manifold if and only if it is a naturally reductive homogeneous space.*
In the sequel, we consider two homogeneous spaces G/H and G'/H' which are T-dual to each other in the sense of the construction given in Sect. 2.6.1 of Chap. 2 and we are going to show that this construction is compatible with 3-symmetry.

As a preparation we recall the construction of the related complex structures. Let us suppose, that $\mathfrak{g}$ is a compact Lie algebra with a subalgebra $\mathfrak{h}$ and that $(M = G/H, g, J)$ is a Riemannian 3-symmetric space with a nearly Kähler structure of above discussed type. Moreover, denote by

$$\mathfrak{g}' = \mathfrak{g}_+ \oplus i\mathfrak{g}_-, \quad \mathfrak{m}' = \mathfrak{m}_+ \oplus i\mathfrak{m}_- \text{ and } \mathfrak{h}' = \mathfrak{h}_+ \oplus i\mathfrak{h}_-$$

the associated decompositions of a fixed T-dual space $M' = G'/H'$.

In this situation, there exists a natural almost complex structure J' on M' which we shortly recall next, cf. Section 3.4 of [82]. Firstly, one decomposes $\mathfrak{gl}(\mathfrak{m})$ into

$$\mathfrak{gl}(\mathfrak{m})_+ := \{A \in \mathfrak{gl}(\mathfrak{m}) \,|\, A(\mathfrak{m}_+) \subset \mathfrak{m}_+, \; A(\mathfrak{m}_-) \subset \mathfrak{m}_-\},$$

$$\mathfrak{gl}(\mathfrak{m})_- := \{B \in \mathfrak{gl}(\mathfrak{m}) \,|\, B(\mathfrak{m}_+) \subset \mathfrak{m}_-, \; B(\mathfrak{m}_-) \subset \mathfrak{m}_+\}.$$

The Lie algebra $\mathfrak{gl}(\mathfrak{m})_+ \oplus i\mathfrak{gl}(\mathfrak{m})_- \subset \mathfrak{gl}(\mathfrak{m})^{\mathbb{C}}$ is isomorphic to $\mathfrak{gl}(\mathfrak{m}')$ via the extension of the following definition

$$A(iX) := iA(X), \; A(Y) := A(Y), \; (iB)(iX) := -B(X), \; (iB)(Y) := iB(Y),$$

where $X \in \mathfrak{m}_-$ and $Y \in \mathfrak{m}_+$ and $A \in \mathfrak{gl}(\mathfrak{m})_+$ and $B \in \mathfrak{gl}(\mathfrak{m})_-$. Further denote by $j \in \mathfrak{gl}(\mathfrak{m})$ the linear map associated to the invariant complex structure J on G/H.

Assume on the one hand, that one even has $j \in \mathfrak{gl}(\mathfrak{m})_+$, then (as shown in Proposition 3.5 of [82]) using the above identification of $\mathfrak{gl}(\mathfrak{m})_+ \oplus i\mathfrak{gl}(\mathfrak{m})_-$ and $\mathfrak{gl}(\mathfrak{m}')$ the map $j \in \mathfrak{gl}(\mathfrak{m})_+ \subset \mathfrak{gl}(\mathfrak{m}')$ induces an invariant almost complex structure J' on G'/H', such that if g is Hermitian for J then J' is pseudo-Hermitian for g'. If on the other hand one has $j \in \mathfrak{gl}(\mathfrak{m})_-$, then $ij \in \mathfrak{gl}(\mathfrak{m})_- \subset \mathfrak{gl}(\mathfrak{m}')$ is a para-complex structure on G'/H'.

For the 3-symmetric case we recover the 3-symmetry using Eq. (3.89), i.e. for $j \in \mathfrak{gl}(\mathfrak{m})_+$ one has

$$\sigma_{|\mathfrak{m}} = s_{*|\mathfrak{m}} = -\frac{1}{2}Id + \frac{\sqrt{3}}{2}j \in \mathfrak{gl}(\mathfrak{m})_+,$$

which induces as before an endomorphism of $\mathfrak{m}'$

$$\sigma_{|\mathfrak{m}'} = -\frac{1}{2}Id + \frac{\sqrt{3}}{2}j \in \mathfrak{gl}(\mathfrak{m})_+ \subset \mathfrak{gl}(\mathfrak{m}'),$$

i.e. after extending σ by the identity on $\mathfrak{h}'$ this yields a local 3-symmetry for G'/H' and assuming G' to be simply connected one may integrates σ to a 3-symmetry of G'.

By construction it follows, that if $(\mathfrak{g}, \mathfrak{m}, \mathfrak{h}, \langle \cdot, \cdot \rangle)$ is naturally reductive, the T-dual $(\mathfrak{g}', \mathfrak{m}', \mathfrak{h}', \langle \cdot, \cdot \rangle')$ is naturally reductive, too. Using Proposition 3.9.15 it follows, that the T-dual of a nearly Kähler 3-symmetric space is nearly pseudo-Kähler. Summarising our discussion we have shown.

Theorem 3.9.16 *Let $(G/H, J, g)$ be a nearly Kähler 3-symmetric space (with compact G) associated to $(\mathfrak{g}, \mathfrak{m}, \mathfrak{h}, \langle \cdot, \cdot \rangle)$ with the above described nearly Kähler* *structure and G'/H' be a T-dual of G/H with data $(\mathfrak{g}', \mathfrak{m}', \mathfrak{h}', \langle \cdot, \cdot \rangle')$ such that the map j associated to the invariant complex structure J lies in $\mathfrak{gl}(\mathfrak{m})_+$, then $(G'/H', J', g')$ is a nearly pseudo-Kähler 3-symmetric space.*

A natural question is starting with some homogeneous nearly Kähler manifold G/H as above to give a classification of all T-dual spaces. Even though there is no general answer to this question the cases of interest for the sequel are discussed in [82]. Let us recall, that by [25] the list of homogenous strict nearly Kähler six-manifolds is

$$S^6 = G_2/SU(3),$$

$$\mathbb{C}P^3 = Sp(2)/(SU(2) \times U(1)),$$

$$\mathbb{F}(1,2) = SU(3)/(U(1) \times U(1)) \text{ and}$$

$$S^3 \times S^3 = (SU(2) \times SU(2) \times SU(2))/\Delta(SU(2)).$$

For these spaces all possible T-duals (given in [82]) are the following **pruefen**

$$S^{2,4} = G_2^*/SU(1,2), \text{cf. Chap. 4 or [82]},$$

$$\mathcal{Z}(S^{2,2}) = SO^+(2,3)/U(1,1), \text{ c.f. Example 3.1 of [82]},$$

$$\mathcal{Z}(S^{4,0}/\mathbb{Z}_2) = SO^+(4,1)/U(2), \text{ c.f. Example 3.1 of [82]},$$

$$\mathcal{Z}(\mathbb{C}P^{2,0}) = SU(2,1)/(U(1) \times U(1)), \text{ c.f. Example 3.2 of [82]},$$

$$\mathcal{Z}((SL(3,\mathbb{R})/GL^+(2,\mathbb{R})) = SL^+(3,\mathbb{R})/\mathbb{R}^* \cdot SO(2), \text{ c.f. Example 3.2 of [82]},$$

$$SL(2,\mathbb{R}) \times SL(2,\mathbb{R}) = (SU(1,1) \times SU(1,1) \times SU(1,1))/\Delta(SU(1,1)).$$

Let us recall, that the twistor spaces already appeared in Sects. 3.3 and 3.4 of the present chapter. Moreover, one may wonders, if one obtains all nearly pseudo-Kähler structures as T-duals of some homogeneous space G/H with a compact Lie group G. The answer can be found in a recent preprint [11], where six-dimensional homogeneous almost complex structures with semi-simple isotropy have been classified. In this list an example of a left invariant nearly pseudo-Kähler structure on a solvable Lie group is given (cf. Remark 3 of [11]), which does not appear in the above list of T-dual spaces.

3.10 Lagrangian Submanifolds in Nearly Pseudo-Kähler Manifolds

This section is based on results with Smoczyk and Schäfer [111] which are extended to pseudo-Riemannian signature in Sects. 3.10.3 and 3.10.4.

3.10.1 Definitions and Geometric Identities

For the rest of this section let us assume that $L \subset M$ is a Lagrangian submanifold[11] of a nearly pseudo-Kähler manifold (M^{2n}, J, g) in the sense of the next definition.

Definition 3.10.1 Let (M^{2n}, J, g, ω) be a nearly pseudo-Kähler manifold. A submanifold $\iota : L^n \to M^{2n}$ is called Lagrangian submanifold, provided that one has $\omega_{|TL} = \iota^* \omega = 0$, the dimension n of L is half the dimension $2n$ of M and that $\iota^* g$ is non-degenerate.

Since $n = \dim(L) = \frac{1}{2} \dim(M)$ we have, that for a Lagrangian submanifold

$$g(JX, Y) = 0, \quad \forall X, Y \in TL \quad \Leftrightarrow \quad J : TL \to T^\perp L \quad \text{is an isomorphism.}$$

We observe that the (symmetric) signature of the metric g restricted to L is (p, q) if the signature of g is $(2p, 2q)$. From $\omega_{|TL} = 0$ we deduce $d\omega_{|TL} = 0$. On the other hand (3.1) implies

$$d\omega(X, Y, Z) = 3g((\nabla_X J)Y, Z).$$

From this and the symmetries of ∇J the following Lemma easily follows (see also [77]).

Lemma 3.10.2 *Suppose $L \subset M$ is a Lagrangian submanifold in a nearly Kähler manifold (M, J, g) with (possibly) indefinite metric. Then*

$$(\nabla_X J)Y \in T^\perp L, \quad \forall X, Y \in TL, \tag{3.101}$$

$$(\nabla_X J)Y \in T^\perp L, \quad \forall X, Y \in T^\perp L, \tag{3.102}$$

$$(\nabla_X J)Y \in TL, \text{ if } X \in TL, Y \in T^\perp L \text{ or if } X \in T^\perp L, Y \in TL. \tag{3.103}$$

Denote by II the second fundamental form of the Lagrangian immersion $L \subset M^{2n}$ into a nearly Kähler manifold M.

Proposition 3.10.3 *For a Lagrangian submanifold $L \subset M^{2n}$ in a nearly Kähler manifold (with possibly indefinite metric) we have the following information.*

(i) *The second fundamental form is given by $\langle II(X, Y), U \rangle = \langle \bar{\nabla}_X Y, U \rangle$ for $X, Y \in \Gamma(TL)$ and $U \in \Gamma(T^\perp L)$.*

(ii) *The tensor $C(X, Y, Z) := \langle II(X, Y), JZ \rangle = \omega(II(X, Y), Z), \forall X, Y, Z \in TL$ is totally symmetric.*

---

[11]In order to compute expressions like for example $\nabla_X Y$ one needs to extend the vector fields on L to vector fields on M. It is common to use the same symbols for the extended vector fields, since the induced objects do not depend on the choice of extension.

Proof From Lemma 3.10.2 we compute for $X, Y \in \Gamma(TL)$ and $U \in \Gamma(T^\perp L)$ the second fundamental form II

$$\langle II(X, Y), U \rangle = \langle \nabla_X Y, U \rangle = \langle \bar\nabla_X Y - \frac{1}{2} J(\nabla_X J)Y, U \rangle = \langle \bar\nabla_X Y, U \rangle .$$

This yields part (i). Next we prove (ii): First we observe for $X, Y, Z \in \Gamma(TL)$

$$C(X, Y, Z) = \langle II(X, Y), JZ \rangle = \langle \bar\nabla_X Y, JZ \rangle = -\langle Y, \bar\nabla_X(JZ) \rangle$$
$$= -\langle Y, J\bar\nabla_X Z \rangle = \langle \bar\nabla_X Z, JY \rangle = C(X, Z, Y).$$

Since the second fundamental form is symmetric, it follows that C is totally symmetric.   □

Next we generalise an identity of [51] to nearly Kähler manifolds of arbitrary dimension and signature of the metric. This and the next lemma will be crucial to prove that Lagrangian submanifolds in strict nearly (pseudo-)Kähler six-manifolds and in twistor spaces Z^{4n+2} over quaternionic Kähler manifolds with their canonical nearly Kähler structure are minimal. A six-dimensional version of the Lemma was also proved in [71], see also Remark 3.10.7.

Lemma 3.10.4 *The second fundamental form II of a Lagrangian immersion $L \subset M^{2n}$ into a nearly (pseudo-)Kähler manifold and the tensor ∇J satisfy the following identity*

$$\langle II(X, J(\nabla_Y J)Z), U \rangle = \langle J(\nabla_{II(X,Y)} J)Z, U \rangle + \langle J(\nabla_Y J)II(X, Z), U \rangle \qquad (3.104)$$

with $X, Y \in TL$ and $U \in T^\perp L$.

Proof The proof of this identity uses $\bar\nabla J = 0$, $\bar\nabla(\nabla J) = 0$ and Lemma 3.10.2. With $X, Y, Z \in \Gamma(TL)$ and $U \in \Gamma(T^\perp L)$ we obtain

$$\langle II(X, J(\nabla_Y J)Z), U \rangle \quad = \quad \langle \bar\nabla_X(J(\nabla_Y J)Z), U \rangle \overset{\bar\nabla J=0}{=} \langle J\bar\nabla_X(\nabla_Y J)Z), U \rangle$$

$$\overset{\bar\nabla((\nabla J))=0}{=} \langle J[(\nabla_{\bar\nabla_X Y} J)Z + (\nabla_Y J)\bar\nabla_X Z], U \rangle$$

$$= \quad \langle J[-(\nabla_Z J)\bar\nabla_X Y + (\nabla_Y J)\bar\nabla_X Z], U \rangle$$

$$= \quad -\langle \bar\nabla_X Y, (\nabla_Z J)JU \rangle + \langle \bar\nabla_X Z, (\nabla_Y J)JU \rangle$$

$$= \quad -\langle II(X, Y), (\nabla_Z J)JU \rangle + \langle II(X, Z), (\nabla_Y J)JU \rangle$$

$$= \quad -\langle J(\nabla_Z J)II(X, Y), U \rangle + \langle J(\nabla_Y J)II(X, Z), U \rangle$$

$$= \quad \langle J(\nabla_{II(X,Y)} J)Z, U \rangle + \langle J(\nabla_Y J)II(X, Z), U \rangle.$$

This is exactly the claim of the Lemma.   □

Given the tensor $C(X, Y, T(Z, V))$ we define the following traces

$$\alpha(X, Y) := \sum_{i=1}^{n} \sigma_i \, C(e_i, X, T(e_i, Y)),$$

$$\beta(X, Y, Z) := \sum_{i=1}^{n} \sigma_i \, C(T(e_i, X), Y, T(e_i, Z)),$$

where $\{e_1, \ldots, e_n\}$ is a local (pseudo-)orthonormal frame of TL and where we set $\sigma_i = g(e_i, e_i)$.

Lemma 3.10.5 *For a Lagrangian immersion in a nearly (pseudo-)Kähler manifold and any $X, Y, Z, V \in TL$ holds:*

$$C(X, Y, T(Z, V)) + C(X, Z, T(V, Y)) + C(X, V, T(Y, Z)) = 0, \quad (3.105)$$

$$\alpha(X, Y) - \alpha(Y, X) = \langle \vec{H}, JT(X, Y) \rangle, \quad (3.106)$$

$$\beta(X, Y, Z) = \beta(Z, Y, X) = \beta(Y, X, Z) + \alpha(T(Y, X), Z), \quad (3.107)$$

$$\alpha(T(X, Y), Z) + \alpha(T(Y, Z), X) + \alpha(T(Z, X), Y) = 0. \quad (3.108)$$

Here $\vec{H}$ denotes the mean curvature vector of L.

Proof Let us first rewrite the identity in Lemma 3.10.4 in terms of the tensor C and the torsion $T(X, Y) = -J(\nabla_X J)Y$ of $\bar{\nabla}$. Let $X, Y, Z, V \in \Gamma(TL)$ be arbitrary. Then Lemma 3.10.4 gives

$$
\begin{aligned}
C(X, T(Z, Y), V) &= \langle II(X, J(\nabla_Y J)Z), JV \rangle \\
&= \langle J(\nabla_{II(X,Y)} J)Z, JV \rangle + \langle J(\nabla_Y J)II(X, Z), JV \rangle \\
&= \langle J(\nabla_Z J)(JV), II(X, Y) \rangle - \langle J(\nabla_Y J)(JV), II(X, Z) \rangle \\
&= -\langle J^2(\nabla_Z J)V, II(X, Y) \rangle + \langle J^2(\nabla_Y J)V, II(X, Z) \rangle \\
&= \langle JT(Z, V), II(X, Y) \rangle - \langle JT(Y, V), II(X, Z) \rangle \\
&= C(X, Y, T(Z, V)) - C(X, Z, T(Y, V)).
\end{aligned}
$$

This is (3.105). Taking a trace gives

$$
\begin{aligned}
\alpha(X, Y) &= \sum_{i=1}^{n} \sigma_i \, C(e_i, X, T(e_i, Y)) \\
&\overset{(3.105)}{=} \sum_{i=1}^{n} \sigma_i \, C(e_i, e_i, T(X, Y)) + \sum_{i=1}^{n} \sigma_i \, C(e_i, Y, T(e_i, X)) \\
&= \langle \vec{H}, JT(X, Y) \rangle + \alpha(Y, X),
\end{aligned}
$$

which is (3.106). The first identity in (3.107) is clear since C is fully symmetric. If we apply (3.105) to $\beta(X, Y, Z)$, then we get

$$\beta(X, Y, Z) = \sum_{i=1}^{n} \sigma_i C(T(e_i, X), Y, T(e_i, Z))$$

$$= -\sum_{i=1}^{n} \sigma_i C(T(X, Y), e_i, T(e_i, Z)) - \sum_{i=1}^{n} \sigma_i C(T(Y, e_i), X, T(e_i, Z))$$

$$= \sum_{i=1}^{n} \sigma_i C(e_i, T(Y, X), T(e_i, Z)) + \sum_{i=1}^{n} \sigma_i C(T(e_i, Y), X, T(e_i, Z))$$

$$= \alpha(T(Y, X), Z) + \beta(Y, X, Z) .$$

This is the second identity in (3.107). In view of this we also get

$$\alpha(T(X, Y), Z) + \alpha(T(Y, Z), X) + \alpha(T(Z, X), Y)$$
$$= \beta(Y, X, Z) - \beta(X, Y, Z) + \beta(Z, Y, X)$$
$$\quad - \beta(Y, Z, X) + \beta(X, Z, Y) - \beta(Z, X, Y)$$
$$= 0$$

and this is (3.108).   □

3.10.2   Lagrangian Submanifolds in Nearly Kähler Six-Manifolds

By a well known theorem of Ejiri [51] Lagrangian submanifolds of S^6 are minimal. In this section we will see that this is a special case of a much more general theorem which is a consequence of Lemma 3.10.4 and was shown independently for Riemannian metrics in Theorem 7 of [71].

Theorem 3.10.6 *Let L^3 be a Lagrangian immersion in a strict nearly (pseudo-)Kähler six-manifold M^6. Then we have*

$$\alpha = 0 , \tag{3.109}$$

$$\vec{H} = 0 . \tag{3.110}$$

In particular, any Lagrangian immersion in a strict nearly (pseudo-)Kähler six-manifold is orientable and minimal.

Proof Let $\{e_1, e_2, e_3\}$ be an orthonormal basis of T_pL for a fixed point $p \in L$. From the skew-symmetry of $\langle T(X, Y), Z \rangle$ we see that there exists a (nonzero) constant a such that $T(e_1, e_2) = a\sigma_3 e_3$. Then we also have $T(e_2, e_3) = a\sigma_1 e_1$, $T(e_3, e_1) = a\sigma_2 e_2$. The symmetry of C implies

$$\alpha(e_1, e_1) = \sigma_1 C(e_1, e_1, T(e_1, e_1)) + \sigma_2 C(e_2, e_1, T(e_2, e_1)) + \sigma_3 C(e_3, e_1, T(e_3, e_1))$$
$$= 0 - a\sigma_2\sigma_3 C(e_2, e_1, e_3) + a\sigma_3\sigma_2 C(e_3, e_1, e_2) = 0$$

and

$$\alpha(e_1, e_2) = \sigma_1 C(e_1, e_1, T(e_1, e_2)) + \sigma_2 C(e_2, e_1, T(e_2, e_2)) + \sigma_3 C(e_3, e_1, T(e_3, e_2))$$
$$= a\sigma_1\sigma_3 C(e_1, e_1, e_3) + 0 - a\sigma_3\sigma_1 C(e_3, e_1, e_1) = 0.$$

Similarly we prove that $\alpha(e_i, e_j) = 0$ for all $i, j = 1, \ldots, 3$. This shows $\alpha = 0$. But then (3.106) also implies $\vec{H} = 0$. The observation, that the frame $\{e_1, e_2, e_3\}$ defines an orientation on L, finishes the proof. The fact that a is a constant was not used in the proof. $\qquad\square$

Remark 3.10.7 The constant a in the formula $T(e_1, e_2) = ae_3$ from above is related to the type constant α of the nearly Kähler manifold M, cf. Sect. 3.1.1 of this chapter, by the formula

$$a^2 = \alpha.$$

A six-dimensional strict nearly Kähler manifold is of constant type and a nearly Kähler manifold of constant type has dimension 6 [67]. The authors of [71] used the Eq. (3.5) to obtain a six-dimensional version of Lemma 3.10.4 for arbitrary nearly Kähler six-manifolds.

In the pseudo-Riemannian case we only have the relation $a^2 = |\alpha|$. The sign of the type constant depends on the signature $(2p, 2q)$ of g by $sign(p - q)$, see for example [82], see also Sect. 3.1.1 of this chapter.
The connection induced on L by $\bar{\nabla}$ is intrinsic in the following sense.

Proposition 3.10.8 *Let L be a Lagrangian submanifold in a strict nearly (pseudo-)Kähler six-manifold. Then the connection $\bar{\nabla}^L$ on L induced by the connection $\bar{\nabla}$ is completely determined by the restriction of g to L.*

Proof We observe that the torsion of $\bar{\nabla}^L$ considered as a three-form is a constant multiple of the volume form $T^L = c\,\mathrm{vol}_g^L$. A metric connection D with prescribed torsion T^D is known to be unique. If the torsion is totally skew-symmetric we can recover it from the formula

$$g(D_X Y, Z) = g(\nabla_X^g Y, Z) + \frac{1}{2}T(X, Y, Z).$$

This finishes the proof, since $T^L = c\,\mathrm{vol}_g^L$ is determined by g. $\qquad\square$

3.10.3 The Splitting Theorem

The following example shows that Theorem 3.10.6 does not extend to eight dimensions:

Example 3.10.9 Let $L' \subset M^6_{SNK}$ be a (minimal) Lagrangian submanifold in a strict nearly Kähler manifold M^6_{SNK} and suppose $\gamma \subset \Sigma$ is a curve on a Riemann surface Σ. Then the Lagrangian submanifold $L := \gamma \times L' \subset M$ in the nearly Kähler manifold $M := \Sigma \times M_{SNK}$ is minimal, if and only if γ is a geodesic in Σ.

In this section we will see that this is basically the only counterexample to Theorem 3.10.6 that occurs in dimension 8. Nearly Kähler manifolds (M, J, g) split locally into a Kähler factor and a strict nearly Kähler factor and under the assumption, that M is complete and simply connected this splitting is global [98], cf. Theorem 3.2.1 of this chapter for the pseudo-Riemannian case. The natural question answered in the following theorem is in which way Lagrangian submanifolds lie in this decomposition.

These facts motivate the next Theorem.

Theorem 3.10.10 *Let M be a nearly Kähler manifold and L be a Lagrangian submanifold. Then M and L decompose locally into products $M = M_K \times M_{SNK}$, $L = L_K \times L_{SNK}$, where M_K is Kähler, M_{SNK} is strict nearly Kähler and $L_K \subset M_K$, $L_{SNK} \subset M_{SNK}$ are both Lagrangian. The dimension of L_K is given by*

$$\dim L_K = \frac{1}{2} \dim \ker(r)$$

Moreover, if the splitting of M is global and L is simply connected, then L decomposes globally as well.

Proof

i) We define

$$K_p := \{X \in T_p M : rX = 0\}, \quad K_p^\perp := \{Y \in T_p M : \langle X, Y \rangle = 0, \forall X \in K_p\}.$$

Because of $\bar{\nabla} r = 0, \bar{\nabla} g = 0$ this defines two orthogonal smooth distributions

$$\mathcal{D}_K := \bigcup_{p \in M} K_p, \quad \mathcal{D}_{SNK} := \bigcup_{p \in M} K_p^\perp$$

on M.

ii) The splitting theorem of de Rham can be applied, see Sect. 3.2.1 of this chapter, to the distributions $\mathcal{D}_K$ and $\mathcal{D}_{SNK}$ and the nearly Kähler manifold (M, J, g) splits (locally) into a Riemannian product

$$(M, J, g) = (M_K, J_K, g_K) \times (M_{SNK}, J_{SNK}, g_{SNK}),$$

where $TM_K = \mathcal{D}_K$, $TM_{SNK} = \mathcal{D}_{SNK}$. Here (M_K, J_K, g_K) is Kähler and $(M_{SNK}, J_{SNK}, g_{SNK})$ is strict nearly Kähler.

iii) Now let $L \subset M = M_K \times M_{SNK}$ be Lagrangian. We prove that r leaves tangent and normal spaces of L invariant. To see this, we fix an adapted local orthonormal frame field $\{e_1, \ldots, e_{2n}\}$ of M such that $e_1, \ldots, e_n$ are tangent to L and $e_{n+1} = Je_1, \ldots, e_{2n} = Je_n$ are normal to L. Since for any three vectors X, Y, Z we have

$$\langle (\nabla_X J)JZ, (\nabla_Y J)JZ \rangle = \langle J(\nabla_X J)Z, J(\nabla_Y J)Z \rangle = \langle (\nabla_X J)Z, (\nabla_Y J)Z \rangle,$$

we obtain

$$\begin{aligned}
\langle rX, Y \rangle &= \sum_{i=1}^{2n} \langle (\nabla_X J)e_i, (\nabla_Y J)e_i \rangle \\
&= \sum_{i=1}^{n} \langle (\nabla_X J)e_i, (\nabla_Y J)e_i \rangle + \sum_{i=1}^{n} \langle (\nabla_X J)Je_i, (\nabla_Y J)Je_i \rangle \\
&= 2 \sum_{i=1}^{n} \langle (\nabla_X J)e_i, (\nabla_Y J)e_i \rangle.
\end{aligned}$$

Now, if $X \in TL$, $Y \in T^\perp L$, then by Lemma 3.10.2 we have

$$(\nabla_X J)e_i \in T^\perp L, \quad (\nabla_Y J)e_i \in TL,$$

so that

$$\langle (\nabla_X J)e_i, (\nabla_Y J)e_i \rangle = 0, \quad \forall\, i = 1, \ldots, n.$$

Further it follows

$$\langle rX, Y \rangle = 2 \sum_{i=1}^{n} \langle (\nabla_X J)e_i, (\nabla_Y J)e_i \rangle = 0.$$

Since this works for any $X \in TL$, $Y \in T^\perp L$ and since r is selfadjoint we conclude

$$r(TL) \subset TL, \quad r(T^\perp L) \subset T^\perp L.$$

At a given point $p \in L$ we may now choose an orthonormal basis $\{f_1, \ldots, f_n\}$ of $T_p L$ that consists of eigenvectors of $r_{|TL}$ considered as an endomorphism of TL. Since $[r, J] = 0$ and L is Lagrangian, the set $\{f_1, \ldots, f_n, Jf_1, \ldots, Jf_n\}$ then also determines an orthonormal eigenbasis of $r \in \text{End}(TM)$. In particular, since J leaves the eigenspaces invariant, $K_p = \ker(r(p))$ and $T_p L$ intersect in a subspace

K_p^L of dimension $\frac{1}{2}\dim(K_p) = \frac{1}{2}\dim(M_K)$. For the same reason $K_p^\perp \cap T_pL$ gives an $\frac{1}{2}\dim(M_{SNK})$-dimensional subspace. The corresponding distributions, denoted by $\mathscr{D}_K^L$ and $\mathscr{D}_{SNK}^L$ are orthogonal and both integrable, since in view of

$$\mathscr{D}_K^L = \mathscr{D}_K \cap TL, \quad \mathscr{D}_{SNK}^L = \mathscr{D}_{SNK} \cap TL$$

they are given by intersections of integrable distributions. We may now apply again the splitting theorem of de Rham to the Lagrangian submanifold. This completes the proof.

$$\square$$

Remark 3.10.11 A more detailed analysis of the proof of the last theorem shows, that the result can be shown in the pseudo-Riemannian setting:

Let (M, J, g) be a nearly pseudo-Kähler manifold. Suppose, that the distribution $\mathcal{K}$ has constant dimension and admits an orthogonal complement, and the kernel of $r_{|L}$ admits an orthogonal complement in TL, then M and L decompose locally into products $M = M_K \times M_{SNK}$, $L = L_K \times L_{SNK}$, where M_K is Kähler, M_{SNK} is strict nearly pseudo-Kähler and $L_K \subset M_K$, $L_{SNK} \subset M_{SNK}$ are both Lagrangian. Moreover, if the splitting of M is global and L is simply connected, then L decomposes globally as well.

Corollary 3.10.12 *If $L \subset M$ is Lagrangian and $p \in L$ a fixed point, then to each eigenvalue λ of the operator r at p there exists a basis $e_1, \ldots, e_k, f_1, \ldots, f_k$ of eigenvectors of $\mathrm{Eig}(\lambda)$ such that $e_1, \ldots, e_k \in T_pL$, $f_1, \ldots, f_k \in T_p^\perp L$. Here, $2k$ denotes the multiplicity of λ.*

Proposition 3.10.13 *Let (M, J, g) be a nearly Kähler manifold and $L \subset M$ be a Lagrangian submanifold. Then TL and $T^\perp L$ are invariant by the Ricci tensor. In particular the spectrum of Ric is compatible with $TL \oplus T^\perp L$.*

Proof Let us recall (cf. [98]) that the Ricci-tensor satisfies $\langle \mathrm{Ric}\, X, Y \rangle = 0$ if X and Y are vector fields in eigenbundles $Eig(\lambda_X)$ and $Eig(\lambda_Y)$ of the tensor r with different eigenvalue $\lambda_X \neq \lambda_Y$. If X, Y belong to the same eigenvalue λ then $\langle \mathrm{Ric}\, X, Y \rangle$ is given by the following formula

$$\langle \mathrm{Ric}\, X, Y \rangle = \frac{\lambda}{4}\langle X, Y \rangle + \frac{1}{\lambda}\sum_{i=1}^{\mu} \lambda_i \langle r^{Eig(\lambda_i)}X, Y \rangle, \tag{3.111}$$

where μ is the number of different eigenvalues of r and $r^{Eig(\lambda_i)}$ is defined by

$$\langle r^{Eig(\lambda_i)}X, Y \rangle = -\mathrm{tr}_{Eig(\lambda_i)}[(\nabla_X J) \circ (\nabla_Y J)].$$

Like in the proof of Theorem 3.10.10 we obtain using Corollary 3.10.12

$$r^{Eig(\lambda_i)}(TL) \subset TL, \quad r^{Eig(\lambda_i)}(T^\perp L) \subset T^\perp L.$$

Equation (3.111) implies that

$$\text{Ric}(TL) \subset TL, \quad \text{Ric}(T^\perp L) \subset T^\perp L.$$

This further implies that the spectrum of Ric is compatible with the decomposition $TL \oplus T^\perp L$ and finishes the proof.   □

Remark 3.10.14 The last proposition gives in the minimal case (only) a partial information on the Ricci curvature of $L \subset M$. Recall the Gauss equation

$$\langle R^L(V, W)X, Y \rangle = \langle R(V, W)X, Y \rangle + \langle II(V, X), II(W, Y) \rangle - \langle II(V, Y), II(W, X) \rangle$$

which implies using minimality

$$\sum_{i=1}^{\dim L} \langle R^L(e_i, W)e_i, Y \rangle = \sum_{i=1}^{\dim L} \langle R(e_i, W)e_i, Y \rangle - \sum_{i=1}^{\dim L} \langle II(e_i, Y), II(e_i, W) \rangle.$$

It is straight-forward to show, that the second term on the right hand-side vanishes if and only if II is zero, i.e. L is a totally geodesic manifold. In that case Proposition 3.10.13 yields the Ricci tensor of L.

Let us recall the situation in dimension 8 and 10 [67, 98].

Proposition 3.10.15

(i) *Let M^8 be a simply connected complete nearly Kähler manifold of dimension 8. Then M^8 is a Riemannian product $M^8 = \Sigma \times M^6_{SNK}$ of a Riemannian surface Σ and a six-dimensional strict nearly Kähler manifold M^6_{SNK}.*
(ii) *Let M^{10} be a simply connected complete nearly Kähler manifold of dimension 10. Then M^{10} is either the product $M^4_K \times M^6_{SNK}$ of a Kähler surface M^4_K and a six-dimensional strict nearly Kähler manifold M^6_{SNK} or M is a twistor space over a positive, eight dimensional quaternionic Kähler manifold.*

Note, that any complete, simply connected eight dimensional positive quaternionic Kähler manifold equals one of the following three spaces: $\mathbb{HP}^2, Gr_2(\mathbb{C}^2), G_2/SO(4)$.

In the next theorem, part (i) and (ii) collect the information on Lagrangian submanifolds in nearly Kähler manifolds of dimension 8 and 10.

Theorem 3.10.16

(i) *Let L be a Lagrangian submanifold in a simply connected nearly Kähler manifold M^8. Then $M^8 = \Sigma \times M^6_{SNK}$, where Σ is a Riemann surface, M^6_{SNK} is strict nearly Kähler and $L = \gamma \times L'$ is a product of a (real) curve $\gamma \subset \Sigma$ and a minimal Lagrangian submanifold $L' \subset M^6_{SNK}$.*
(ii) *Let L be a Lagrangian submanifold in a simply connected complete nearly Kähler manifold M^{10}, then either*

(a) $M^{10} = M_K^4 \times M_{SNK}^6$ and the manifold $L = S \times L'$ is a product of a Lagrangian (real) surface $S \subset M_K^4$ and a minimal Lagrangian submanifold $L' \subset M_{SNK}^6$ or

(b) the manifold L is a Lagrangian submanifold in a twistor space over a positive, eight dimensional quaternionic Kähler manifold.

(iii) Let M_1, M_2 be two nearly Kähler manifolds. Denote the operator r on M_i by r_i, $i = 1, 2$. If $\mathrm{Spec}(r_1) \cap \mathrm{Spec}(r_2) = \emptyset$ and $L \subset M_1 \times M_2$ is Lagrangian, then L splits (locally) into $L = L_1 \times L_2$, where $L_i \subset M_i$, $i = 1, 2$ are Lagrangian. If L is simply connected, then the decomposition is global.

Proof This is a combination of the results of Theorem 3.10.10, Corollary 3.10.12 and Proposition 3.10.15.   □

Remark 3.10.17 As in Remark 3.10.11 we may note, that using our results one can generalise Theorem 3.10.16 to the case of indefinite metrics, where we omit part (iii):

Theorem 3.10.18

(i) Let L be a Lagrangian submanifold in a simply connected nice nearly pseudo-Kähler manifold M^8. Then $M^8 = \Sigma \times M_{SNK}^6$, where Σ is a Riemann surface,[12] M_{SNK}^6 is strict nearly Kähler and $L = \gamma \times L'$ is a product of a (real) curve $\gamma \subset \Sigma$ and a minimal Lagrangian submanifold $L' \subset M_{SNK}^6$.

(ii) Let L be a Lagrangian submanifold in a simply connected complete nice decomposable nearly pseudo-Kähler manifold M^{10}, such that the kernel[13] of $r_{|TL}$ admits an orthogonal complement, then either

(a) $M^{10} = M_K^4 \times M_{SNK}^6$ and the manifold $L = S \times L'$ is a product of a Lagrangian (real) surface $S \subset M_K^4$ and a minimal Lagrangian submanifold $L' \subset M_{SNK}^6$ or

(b) the manifold L is a Lagrangian submanifold in a twistor space over a negative, eight dimensional quaternionic Kähler manifold or a para-quaternionic Kähler manifold.

Theorems 3.10.16 (3.10.16) and 3.10.18 (3.10.18), parts (b) motivate the discussion of Lagrangian submanifolds in twistor spaces in the subsequent section. Indeed, the results derived in the next section imply that Lagrangian submanifolds in twistor spaces are, regardless their dimension, always minimal.

---

[12]Remark, that the restriction of g to Σ is always definite.

[13]Let us recall, that in the case (b) r has trivial kernel.

3.10.4 Lagrangian Submanifolds in Twistor Spaces

An important class of examples for nearly pseudo-Kähler manifolds is given by twistor spaces Z^{4n+2} over quaternionic Kähler or para-quaternionic Kähler manifolds N^{4n}, as we have seen in Sects. 3.3 and 3.4 of this chapter.

For the readers convenience, let us shortly recall that the twistor space is the bundle of almost complex structures in the quaternionic bundle Q over the (para-)quaternionic Kähler manifold N. It can be endowed with a Kähler structure (Z, J^Z, g^Z), such that the projection $\pi : Z \rightarrow N$ is a Riemannian submersion with totally geodesic fibres S^2. Denote by $\mathcal{H}$ and $\mathcal{V}$ the horizontal and the vertical distributions of the submersion π. Then the direct sum decomposition

$$TZ = \mathcal{H} \oplus \mathcal{V} \qquad (3.112)$$

is orthogonal and compatible with the complex structure J^Z. Let us consider now a second almost Hermitian structure (J, g) on Z which is defined by

$$g := \begin{cases} g^Z(X, Y), & \text{for } X, Y \in \mathcal{H}, \\ \frac{1}{2} g^Z(V, W), & \text{for } V, W \in \mathcal{V}, \\ g^Z(V, X) = 0, & \text{for } V \in \mathcal{V}, X \in \mathcal{H} \end{cases}$$

and

$$J := \begin{cases} J^Z \text{ on } \mathcal{H}, \\ -J^Z \text{ on } \mathcal{V}. \end{cases}$$

Note, that in view of (3.112), the decomposition $TZ = \mathcal{H} \oplus \mathcal{V}$ is also compatible w.r.t. J and orthogonal w.r.t. g.

The manifold (Z, J, g) is a strict nearly pseudo-Kähler manifold and the distributions $\mathcal{V}$ and $\mathcal{H}$ are parallel w.r.t. the connection $\bar{\nabla}$. The projection π is also a Riemannian submersion with totally geodesic fibres for the metric g.

Let us summarise some information which will be useful later in this section.

Lemma 3.10.19 *In this situation we have the following information:*

(a) *The torsion* $T = -J\nabla J$ *of the characteristic connection satisfies (see Lemma 3.4.3)*

$$T(X, Y) \in \mathcal{V}, \quad \text{for } X, Y \in \mathcal{H}, \qquad (3.113)$$

$$T(X, V) \in \mathcal{H}, \quad \text{for } X \in \mathcal{H}, V \in \mathcal{V}, \qquad (3.114)$$

$$T(U, V) = 0, \quad \text{for } U, V \in \mathcal{V}. \qquad (3.115)$$

(b) *The association (see Lemma 3.4.3)*

$$\mathcal{H} \ni X \mapsto T(Y, X) \in \mathcal{V} \tag{3.116}$$

is surjective for $0 \neq Y \in \mathcal{H}$ *and the map*

$$\Phi^V : \mathcal{H} \ni X \mapsto T(V, X) \tag{3.117}$$

with $0 \neq V \in \mathcal{V}$ *is invertible and squares to* $\varepsilon k^2 Id_{\mathcal{H}}$ *for some* $k \in \mathbb{R}, \varepsilon \in \{\pm 1\}$, *cf. Lemma 3.4.9.*

(c) *The operator* r *has eigenvalues* $\lambda_{\mathcal{H}} = 4k^2$, $\lambda_{\mathcal{V}} = \frac{n-1}{2}\varepsilon\lambda_{\mathcal{H}}$. *If* $n > 1$, *then the eigenbundle of* $\lambda_{\mathcal{H}}$ *is* $\mathcal{H}$ *and* $\mathcal{V}$ *is the eigenbundle of* $\lambda_{\mathcal{V}}$, *cf. Corollary 3.4.17.*

In the rest of this section we consider a nearly pseudo-Kähler manifold $(M = Z, J, g)$ of twistor type and study Lagrangian submanifolds $L \subset M$.

Remark 3.10.20 As will be shown in the next theorem, for $n > 1$ we have

$$\pi^{\mathcal{H}}(TL) = \mathcal{H} \cap TL, \quad \pi^{\mathcal{V}}(TL) = \mathcal{V} \cap TL,$$

where $\pi^{\mathcal{H}}(TL), \pi^{\mathcal{H}}(T^\perp L)$ and $\pi^{\mathcal{V}}(TL), \pi^{\mathcal{V}}(T^\perp L)$ are the orthogonal projections of TL and $T^\perp L$ w.r.t. $\mathcal{H} \oplus \mathcal{V}$.

Lemma 3.10.21 *Let* $L^{2n+1} \subset M^{4n+2}$ *with* $n > 1$ *be a Lagrangian submanifold in a twistor space as described above. Then the second fundamental form II satisfies*

$$II(X, Y) \in \pi^{\mathcal{H}}(T^\perp L), \quad \text{for } X, Y \in \pi^{\mathcal{H}}(TL), \tag{3.118}$$

$$II(X, Y) \in \pi^{\mathcal{V}}(T^\perp L), \quad \text{for } X, Y \in \pi^{\mathcal{V}}(TL), \tag{3.119}$$

$$II(X, Y) = 0, \quad \text{for } X \in \pi^{\mathcal{H}}(TL), Y \in \pi^{\mathcal{V}}(TL). \tag{3.120}$$

Proof The second fundamental form is given by $C(X, Y, Z) = \langle \bar{\nabla}_X Y, JZ \rangle$ for $X, Y, Z \in \Gamma(TL)$. The lemma follows since the decomposition (3.112) is $\bar{\nabla}$-parallel, orthogonal and J-invariant and as the tensor C is completely symmetric. $\quad\square$
With these preparations we prove the next result.

Theorem 3.10.22 *Let* $L^{2n+1} \subset M^{4n+2}$ *be a Lagrangian submanifold in a nearly pseudo-Kähler manifold of the above type. Then L is minimal. If* $n > 1$, *then the tangent space of L splits into a one-dimensional vertical part and a 2n-dimensional horizontal part. Moreover, the second fundamental form II of the vertical normal direction vanishes completely if* $n > 1$.

Proof

i) By Theorem 3.10.6 it suffices to consider the case $n > 1$.
ii) Let $L \subset M$ be a Lagrangian submanifold. Since $n > 1$, the two eigenvalues $\lambda_{\mathcal{H}}$, $\lambda_{\mathcal{V}}$ of r are distinct and the eigenspace $\mathcal{V}$ of $\lambda_{\mathcal{V}}$ is two-dimensional. By Corollary 3.10.12 this induces a one-dimensional vertical tangential distribution

$\mathcal{D}$ on L in the Riemannian case. In the pseudo-Riemannian case this follows, since the restriction of the metric to $\mathcal{V}$ is definite. In particular, $\mathcal{D}$ is not isotropic. Then, by the Lagrangian condition, we get $\mathcal{D} := \pi^{\mathcal{V}}(TL)$.

iii) Denote by $\mathcal{D}^{\perp}$ the orthogonal complement of $\mathcal{D}$ in TL. We claim, that the trace of the second fundamental form II of L restricted to $\mathcal{D}^{\perp}$ is zero.

Proof First we observe that by Lemma 3.10.21 we can restrict the second fundamental form II to $\mathcal{D}^{\perp} = \pi^{\mathcal{H}}(TL)$. We fix $U \in \mathcal{D}$ of unit length. Using Lemmas 3.10.2 and 3.10.19 we observe, that $\Phi(X) := \frac{1}{k}J(\nabla_U J)X$ defines an (almost) (para-)complex structure on $\mathcal{D}^{\perp}$ which is compatible with the metric. With $X \in \mathcal{D}^{\perp}$ and $\Phi^2 = \varepsilon Id$ we compute

$$II(X, X) = \varepsilon II(X, \Phi(\Phi(X))) = \varepsilon \frac{1}{k} II(X, J(\nabla_U J) \Phi(X))$$

$$= \varepsilon \frac{1}{k} J \left[(\nabla_{II(X,U)} J) \Phi(X) + (\nabla_U J) II(X, \Phi(X)) \right]$$

$$= \varepsilon \frac{1}{k} J(\nabla_U J) II(X, \Phi(X)) = \varepsilon \Phi II(X, \Phi(X)).$$

After polarising we obtain

$$II(\Phi X, \Phi Y) = \varepsilon II(X, Y), \quad \forall X, Y \in \mathcal{D}^{\perp}. \tag{3.121}$$

In particular, taking a trace over (3.121) we get

$$\mathrm{tr}^{\mathcal{D}^{\perp}} II = 0,$$

where we keep in mind, that it holds $g(\Phi\cdot, \Phi\cdot) = -\varepsilon g(\cdot, \cdot)$.

iv) We have

$$\alpha(X, Y) = 0, \quad \forall X, Y \in \mathcal{D}^{\perp}.$$

Proof By (ii) we may choose a pseudo-orthonormal frame $\{e_1, \ldots, e_{2n+1}\}$ of TL such that $e_1, \ldots, e_{2n} \in \mathcal{D}^{\perp}$ and $e_{2n+1} \in \mathcal{D}$. Since

$$\alpha(X, Y) = \sum_{i=1}^{2n+1} \sigma_i C(e_i, X, T(e_i, Y))$$

and the tensor C is fully symmetric we see that by Lemma 3.10.21 all terms on the RHS vanish since either $e_i \in \mathcal{D} = \pi^{\mathcal{V}}(TL)$, $X \in \mathcal{D}^{\perp} = \pi^{\mathcal{H}}(TL)$ or $e_i, X \in \mathcal{D}^{\perp}$ and $T(e_i, Y) \in \mathcal{D}$ (cf. Lemma 3.10.19).

v) By Lemma 3.10.21 and (iii) the mean curvature vector $\vec{H}$ satisfies $J\vec{H} \in \mathcal{D}$. From (3.106) and (iv) we get

$$\langle J\vec{H}, T(X, Y) \rangle = 0, \quad \forall X, Y \in \mathcal{D}^{\perp}. \tag{3.122}$$

Since J maps $\mathcal{V}$ to itself, we also have $J\vec{H} \in \mathcal{D} \subset \mathcal{V}$. Now we choose $X \in \mathcal{D}^{\perp}$ and $\tilde{Y} \in \mathcal{H}$ with

$$T(X, \tilde{Y}) = J\vec{H} \, .$$

This is possible since the map $\mathcal{H} \ni \tilde{Y} \mapsto T(X, \tilde{Y}) \in \mathcal{V}$ is surjective by (3.116). Let $\tilde{Y} = Y + Y^{\perp}$ be the orthogonal decomposition of $\tilde{Y}$ into the tangent and normal parts of $\tilde{Y}$. Note that $Y, Y^{\perp}$ are both horizontal. We claim

$$T(X, Y^{\perp}) = 0.$$

This follows, since $T(X, \cdot)$ maps tangent to tangent and normal to normal vectors and one has

$$T(X, \tilde{Y}) = T(X, Y) + T(X, Y^{\perp}) = J\vec{H} \in TL.$$

Therefore there exist two tangent vectors $X, Y \in \mathcal{D}^{\perp}$ with

$$T(X, Y) = J\vec{H} \, .$$

This implies

$$|\vec{H}|^2 = \langle J\vec{H}, T(X, Y)\rangle \overset{(3.122)}{=} 0,$$

which proves that the mean curvature vector vanishes, as the metric restricted to $\mathcal{V}$ is definite (even in the pseudo-Riemannian case). From this, the fact that $\mathcal{D}$ is one-dimensional and from Lemma 3.10.21 it follows that $II(V, \cdot) = 0$ for any $V \in \mathcal{D}$.

$\square$

Corollary 3.10.23 *Let $L \subset M$ be a Lagrangian submanifold in a twistor space M^{4n+2} as above with $n > 1$. Then the integral manifolds c of the distribution $\mathcal{D}$ are geodesics (hence locally great circles) in the totally geodesic fibres S^2.*

Proof The last theorem implies that the geodesic curvature vanishes and that in consequence an integral manifold c of $\mathcal{D}$ is totally geodesic in the fibres. $\square$

Remark 3.10.24

(i) It is well-known, that the twistor space of $\mathbb{H}P^n$ is $\mathbb{C}P^{2n+1}$. Therefore the above result applies to $(\mathbb{C}P^{2n+1}, J, g)$ endowed with its canonical nearly Kähler structure.

(ii) Using Remark 3.10.14 (for Riemannian metrics) and Lemma 3.10.19 (c) we observe that totally geodesic Lagrangian submanifolds in twistor spaces have two different Ricci eigenvalues with multiplicities $2n$ and 1.

3.10.5 Deformations of Lagrangian Submanifolds in Nearly Kähler Manifolds

Our aim in this section is to study the space of deformations of a given Lagrangian (and hence minimal Lagrangian) submanifold L in a strict six-dimensional nearly (pseudo-)Kähler manifold M^6. In an article by Moroianu et al. [96] the deformation space of nearly Kähler structures on six-dimensional nearly Kähler manifolds has been related to the space of coclosed eigenforms of the Hodge-Laplacian. As we will show below, a similar statement holds for the deformation of Lagrangian submanifolds in strict nearly (pseudo-)Kähler six-manifolds. To this end we assume that

$$F : L \times (-\epsilon, \epsilon) \to M$$

is a smooth variation of Lagrangian immersions $F_t := F(\cdot, t) : L \to M, t \in (-\epsilon, \epsilon)$ into a nearly (pseudo-)Kähler manifold M. Let

$$V := \frac{d}{dt} F_t$$

denote the variation vector field. Since tangential deformations correspond to diffeomorphisms acting on L, we may assume w.l.o.g. that $V \in \Gamma(T^\perp L)$ is a normal vector field. The Cartan formula and $F_t^* \omega = 0$ for all t then imply that

$$0 = d(i_V \omega) + i_V d\omega$$

holds everywhere on L. By the nearly Kähler condition this is equivalent to

$$d(V \lrcorner \omega) + 3 V \lrcorner \nabla \omega = 0 \tag{3.123}$$

on L. Let us define the variation 1-form $\theta \in \Omega^1(L)$ by

$$\theta := V \lrcorner \omega .$$

This Theorem has recently been used in [97]. In this paper the authors relate generalised Killing spinors on spheres to Lagrangian graphs in the nearly Kähler manifold $S^3 \times S^3$.

Theorem 3.10.25 *Let $F_t : L \to M$ be a variation of Lagrangian immersions in a six-dimensional nearly (pseudo-)Kähler manifold M. Then the variation 1-form θ is a coclosed eigenform of the Hodge-Laplacian, where the eigenvalue λ satisfies $\lambda = 9\alpha$ with the type constant α of M. If the metric is positive definite this space is finite dimensional.*

In the case of Riemannian metrics a similar result was also shown in Theorem 7 of [71]. For a more recent study of deformations of Lagrangian submanifolds we refer to [91].

Proof For $X, Y \in TL$ and $V \in T^\perp L$ we compute

$$
\begin{aligned}
(V \lrcorner \nabla \omega)(X, Y) &= \nabla \omega(V, X, Y) \\
&= \langle (\nabla_X J) Y, V \rangle \\
&= \langle J(\nabla_X J) Y, JV \rangle \\
&= -\langle JV, T(X, Y) \rangle \, .
\end{aligned}
$$

Since T induces an orientation on the Lagrangian submanifold by the three-form

$$
\tau(X, Y, Z) := \langle T(X, Y), Z \rangle \, ,
$$

we obtain a naturally defined $*$-operator $* : \Omega^p(L) \to \Omega^{3-p}(L)$ which for 1-forms is given by

$$
*\phi := \frac{\sigma}{\sqrt{|\alpha|}} \, \phi \circ T \, .
$$

Here, α is the type constant of M (cf. Remark 3.10.7) and $\sigma \in \{\pm 1\}$ depends only on the signature. This implies that equation (3.123) can be rewritten in the form

$$
d\theta = 3\sigma \sqrt{|\alpha|} *\theta \, . \tag{3.124}
$$

Consequently, if the signature of the metric g restricted to L is (p, q), we obtain

$$
\mathrm{sign}(p - q) \, \delta\theta = *d*\theta = 0
$$

and

$$
\begin{aligned}
\mathrm{sign}(p - q) \, \delta d\theta \quad &= \quad 3\sigma \sqrt{|\alpha|} *d**\theta \\
&\overset{(*^2=\mathrm{Id})}{=} \quad 3\sigma \sqrt{|\alpha|} *d\theta \\
&\overset{(3.124)}{=} \quad 9|\alpha| * *\theta \\
&\overset{(*^2=\mathrm{Id})}{=} \quad 9|\alpha|\theta \, .
\end{aligned}
$$

In total as the sign of α is also sign$(p - q)$ we get

$$\Delta_{\text{Hodge}}\theta = (\delta d + d\delta)\theta = 9\alpha\theta .$$

This proves the theorem. Since one has $Ric = 5\alpha g$ this is equivalent to

$$\Delta_{\text{Hodge}}\theta = \frac{3}{10} scal\,\theta ,$$

where $scal$ is the scalar curvature of M.   □

Chapter 4
Hitchin's Flow Equations

This chapter relies on work with Leistner et al. [46].

4.1 Half-Flat Structures and Parallel $G_2^{(*)}$-Structures

Now we want to put the algebraic structures considered in the first chapter onto smooth manifolds. This is best done in terms of reductions of the bundle of frames of the manifold. This bundle has $GL(n, \mathbb{R})$ as its structure group if n is the dimension of the manifold. A subbundle whose structure group is a subgroup G of $GL(n, \mathbb{R})$ is a called a reduction of the frame bundle, or a *G-structure*. For example, if $G \subset O(p, q)$, for $p + q = n$, the reduction determines a pseudo-Riemannian metric of signature (p, q) and the distinguished frames are orthonormal with respect to this metric. If $G \subset O(p, q)$, again with $p + q = n$, then a G-structure is called *parallel* if the G-subbundle is invariant under the parallel transport defined by the Levi-Civita connection of the corresponding metric. This is equivalent to the property that the holonomy group of the Levi-Civita is contained in G.

In the following we will consider G-structures that are given by the groups described in the previous sections. According to the notations given there, we denote by $H^{\varepsilon,\tau}$ a real form of $SL(3, \mathbb{C})$ and $G^{\varepsilon,\tau}$ the corresponding real form of $G_2^{\mathbb{C}}$ in which $H^{\varepsilon,\tau}$ is embedded, i.e. $H^{-1,1} = SU(3) \subset SO(6)$, $H^{-1,-1} = SU(1, 2) \subset SO(2, 4)$, $H^{1,1} = SL(3, \mathbb{R}) \subset SO(3, 3)$, $G^{-1,1} = G_2 \subset SO(7)$, and $G^{-1,-1} = G^{1,1} = G_2^* \subset SO(3, 4)$. We will also use the notation $G_2^{(*)}$ as a shorthand for "G_2 respectively G_2^*".

An $H^{\varepsilon,\tau}$-structure is equivalent to a pair of everywhere stable forms $\omega \in \Omega^2 M$ and $\rho \in \Omega^3 M$ on M, considered up to rescaling of ρ by a nonzero constant, that satisfy the compatibility condition

$$\rho \wedge \omega = 0 \tag{4.1}$$

© Springer International Publishing AG 2017
L. Schäfer, *Nearly Pseudo-Kähler Manifolds and Related Special Holonomies*,
Lecture Notes in Mathematics 2201, DOI 10.1007/978-3-319-65807-0_4

corresponding to (2.16) and in addition

$$\phi(\rho) = c \, \phi(\omega) \,, \qquad \text{i.e. } J_\rho^* \rho \wedge \rho = \frac{1}{3} c \, \omega^3 \,, \tag{4.2}$$

for a positive real constant c. Indeed, if an $H^{\varepsilon,\tau}$-structure is given, these forms are obtained by applying the formulae (2.7) and (2.20) to one of the frames of the $H^{\varepsilon,\tau}$-structure. By construction, the stable forms then satisfy (4.2) with $c = 2$.

On the other hand, if $\omega \in \Omega^2 M$ and $\rho \in \Omega^3 M$ are everywhere stable and satisfy (4.1) and (4.2), we can find a local frame, in which they are in normal form after rescaling ρ by a constant. This frame then determines the $H^{\varepsilon,\tau}$-structure.

Note that stable forms define an $H^{\varepsilon,\tau}$-structure, even if they only satisfy (4.1) but not the second compatibility condition (4.2). In this case ρ can always be rescaled by a smooth function such that (4.2) holds. When we say that the pair of stable forms defines an $H^{\varepsilon,\tau}$-structure, we will always assume that *both* compatibility conditions are satisfied. We will call the $H^{\varepsilon,\tau}$-structure *normalised* if $c = 2$. This seems to be a common normalisation for SU(3)-structures in the literature.

Furthermore, one can show that the $H^{\varepsilon,\tau}$-structure is parallel if and only if ρ, $\hat\rho$, and ω are closed. The proof of this fact given in [75, p. 567] generalises to SU(1, 2)-structures and also to SL(3, $\mathbb{R}$)-structures, in the latter case using Frobenius' Theorem instead of the Newlander-Nirenberg Theorem. In all cases the parallel $H^{\varepsilon,\tau}$-structure is equivalent to M being a Ricci-flat (para-)Kähler manifold.

Now we consider a weaker condition, that will turn out to be related to parallel $G_2^{(*)}$-structures.

Definition 4.1.1 An $H^{\varepsilon,\tau}$-structure (ρ, ω) is called *half-flat* if

$$d\rho = 0, \tag{4.3}$$

$$d\sigma = 0, \tag{4.4}$$

where $2\sigma = \omega^2$.

Similarly, a smooth seven-manifold admits a $G_2^{(*)}$-structure if and only if there is a stable three-form φ. Again, this structure is parallel if and only if φ is closed and co-closed, i.e. $d\varphi = d * \varphi = 0$, where $*$ denotes the Hodge operator with respect to the metric induced by the $G_2^{(*)}$-structure. For a proof in both cases see [64, Theorem 4.1].

Note that any orientable hypersurface in a manifold with G_2- or G_2^*-structure admits an $H^{\varepsilon,\tau}$-structure by the algebraic construction described in Proposition 2.1.14. If the $G_2^{(*)}$-structure φ is parallel, the induced $H^{\varepsilon,\tau}$-structure is half-flat due to Eqs. (2.23) and (2.27). For the various results on the SU(3)-structures on hypersurfaces in G_2-structures, we refer to [27, 28] and references therein.

On the other hand, certain one-parameter families of half-flat structures define parallel $G_2^{(*)}$-structures.

Proposition 4.1.2 *Let $H^{\varepsilon,\tau}$ be a real form of* $SL(3, \mathbb{C})$, *$G^{\varepsilon,\tau}$ the corresponding real form of* $G_2^{\mathbb{C}}$ *and* (ρ, ω) *a one-parameter family of $H^{\varepsilon,\tau}$-structures on a six-manifold M with a parameter t from an interval I. Then, the three-form*

$$\varphi = \omega \wedge dt + \rho$$

defines a parallel $G^{\varepsilon,\tau}$-structure on $M \times I$ if and only if the $H^{\varepsilon,\tau}$-structure (ρ, ω) is half-flat for all t and satisfies the following evolution equations

$$\dot{\rho} = d\omega, \tag{4.5}$$

$$\dot{\sigma} = d\hat{\rho} \tag{4.6}$$

with $\sigma = \frac{1}{2}\omega^2$.

Proof Let (ρ, ω) be an $H^{\varepsilon,\tau}$-structure and $\varphi = \omega \wedge dt + \rho$ a stable three-form on $\check{M} := M \times I$. By (2.27), the Hodge-dual of φ is given by

$$*\varphi = \varepsilon \left(\hat{\rho} \wedge dt - \sigma \right).$$

Denoting by $\check{d}$ the differential on $\check{M}$ and by d the differential on M we calculate

$$\check{d}\varphi = d\omega \wedge dt + dt \wedge \dot{\rho} + d\rho = (d\omega - \dot{\rho}) \wedge dt + d\rho, \tag{4.7}$$

$$\check{d} * \varphi = \varepsilon \left(d\hat{\rho} \wedge dt - dt \wedge \dot{\sigma} - d\sigma \right) = \varepsilon(d\hat{\rho} - \dot{\sigma}) \wedge dt - \varepsilon d\sigma. \tag{4.8}$$

Thus, φ defines a parallel $G^{\varepsilon,\tau}$-structure if and only if the evolution equations (4.5) and (4.6) and the half-flat equations are satisfied.   □

The evolution equations (4.5) and (4.6) are the *Hitchin flow equations*, as found in [76] for SU(3)-structures, applied to $H^{\varepsilon,\tau}$-structures. Their solutions (ρ, ω), called *Hitchin flow*, have to satisfy possibly dependent conditions in order to yield a parallel $G_2^{(*)}$-structure: the evolution equations and the compatibility equations for the family of half-flat structures. The following theorem shows that the evolution equations together with an initial condition already ensure that the family consists of half-flat structures. A special version of this theorem was proved in [76] under the assumption that M is compact and that $H = SU(3)$.

Theorem 4.1.3 *Let (ρ_0, ω_0) be a half-flat $H^{\varepsilon,\tau}$-structure on a six-manifold M. Furthermore, let $(\rho, \omega) \in \Omega^3 M \times \Omega^2 M$ be a one-parameter family of stable forms with parameters from an interval I satisfying the evolution equations (4.5) and (4.6). If $(\rho(t_0), \omega(t_0)) = (\rho_0, \omega_0)$ for a $t_0 \in I$, then (ρ, ω) is a family of half-flat $H^{\varepsilon,\tau}$-structures. In particular, the three-form*

$$\varphi = \omega \wedge dt + \rho \tag{4.9}$$

defines a parallel $G^{\varepsilon,\tau}$-structure on $M \times I$ and the induced metric

$$g_\varphi = g(t) - \varepsilon dt^2, \tag{4.10}$$

has holonomy contained in $G^{\varepsilon,\tau}$, where $g = g(t)$ is the family of metrics on M associated to (ρ, ω).

Proof Differentiating the evolution equations (4.5) and (4.6) gives $d\dot{\rho} = d\dot{\sigma} = 0$. The initial condition for t_0 was that (ρ_0, ω_0) is half-flat. This implies

$$d\rho = 0,$$

$$d\sigma = 0$$

for all $t \in I$. Hence, in order to obtain a family of half-flat structures we have to verify that the compatibility condition (4.1) holds for all $t \in I$.

Lemma 4.1.4 *Let M be a six-manifold with $H^{\varepsilon,\tau}$-structure (ρ, ω), $\phi : \Omega^3 M \to \Omega^6 M$ defined pointwise by the map $\phi : \Lambda^3 T_p^* M \to \Lambda^6 T_p^* M$ given in Proposition 2.1.4 and $\hat{\rho}$ defined by $d\phi_\rho(\xi) = \hat{\rho} \wedge \xi$ for all $\xi \in \Omega^3 M$. If $\mathcal{L}_X$ denotes the Lie derivative, then*

$$\mathcal{L}_X(\phi(\rho)) = \hat{\rho} \wedge \mathcal{L}_X \rho.$$

Proof First note that the $\mathrm{GL}(n, \mathbb{R})$-equivariance of the map $\phi : \Lambda^3 T_p^* M \to \Lambda^6 T_p^* M$ implies that the corresponding map $\phi : \Omega^3 M \to \Omega^6 M$ is equivariant under diffeomorphisms. Indeed, if ψ is a (local) diffeomorphism of M we get that

$$\psi^*(\phi(\rho)) = \phi(\psi^* \rho).$$

Let ψ_t be the flow of the vector field X. Then the Lie derivative is given by

$$\mathcal{L}_X(\phi(\rho)) = \frac{d}{dt}\left(\psi_t^* \phi(\rho)\right)|_{t=0} = \frac{d}{dt}\phi(\psi_t^* \rho)|_{t=0} = d\phi_\rho(\mathcal{L}_X \rho),$$

implying the statement. $\qquad \qquad \square$

Lemma 4.1.5 *A stable three-form $\rho \in \Omega^3 M$ on a six-manifold satisfies for any $X \in \mathfrak{X}(M)$*

$$\dot{\rho}_X \wedge \rho = -\hat{\rho} \wedge \rho_X, \tag{4.11}$$

$$(d\hat{\rho})_X \wedge \rho = \hat{\rho} \wedge (d\rho)_X, \tag{4.12}$$

where ρ_X denotes the interior product of X with the form ρ.

Proof In order to verify the first identity, we can assume that $\rho = \rho_p$ is a stable three-form on $V = T_p M$ and $X \in V$ for a $p \in M$. If $\lambda(\rho) < 0$, the stabiliser $\mathrm{SL}(3, \mathbb{C})$

of ρ in $GL^+(V)$ acts transitively on $V \setminus \{0\}$. If $\lambda(\rho) > 0$, we can decompose V in the ± 1-eigenspaces $V^\pm$ of J_ρ. The stabiliser $SL(3, \mathbb{R}) \times SL(3, \mathbb{R})$ of ρ in $GL^+(V)$ acts transitively on the dense open subset $V^+ \setminus \{0\} \times V^- \setminus \{0\} \subset V$ and there is an automorphism exchanging V^+ and V^- which stabilises ρ. Thus, it suffices to verify the first identity for the normal form (2.7), (2.8) and $X = e_1$, which is easy.

For the second identity, using Lemma 4.1.4 in the second step, we compute

$$
\begin{aligned}
(d\hat{\rho})_X \wedge \rho - \hat{\rho} \wedge (d\rho)_X &= -d\hat{\rho} \wedge \rho_X + \hat{\rho} \wedge d(\rho_X) - \hat{\rho} \wedge \mathcal{L}_X \rho \\
&= -d(\hat{\rho} \wedge \rho_X) - \mathcal{L}_X(\phi(\rho)) \\
&= -d(\hat{\rho} \wedge \rho_X + \phi(\rho)_X) \\
&\overset{(2.2)}{=} -\frac{1}{2} d(\hat{\rho} \wedge \rho_X + \hat{\rho}_X \wedge \rho).
\end{aligned}
$$

Hence, the first identity (4.11) implies (4.12). $\square$

Using this lemma, we calculate the t-derivative of the six-form $\omega_X \wedge \omega \wedge \rho = \sigma_X \wedge \rho$ for any vector field X:

$$
\begin{aligned}
\frac{\partial}{\partial t}(\sigma_X \wedge \rho) &= \dot{\sigma}_X \wedge \rho + \sigma_X \wedge \dot{\rho} \\
&\overset{(4.5),(4.6)}{=} (d\hat{\rho})_X \wedge \rho + \sigma_X \wedge d\omega \\
&\overset{(4.12)}{=} \hat{\rho} \wedge (d\rho)_X + \omega_X \wedge \omega \wedge d\omega \\
&\overset{(4.3),(4.4)}{=} 0.
\end{aligned}
$$

Together with the initial condition $\omega_0 \wedge \rho_0 = 0$ this implies that $\sigma_X \wedge \rho = 0$ for all $t \in I$ and for all vector fields X. Since ω is nondegenerate, the product of any one-form with $\omega \wedge \rho$ vanishes and thus, the compatibility condition $\omega \wedge \rho = 0$ holds for all t.

The preservation of the normalisation (4.2) in time is shown in [76], in the final part of the proof of Theorem 8. The idea is to compute the second derivative of the volume form assigned to a stable three-form. In fact, the proof holds literally for all signatures since all it uses is the first compatibility condition we have just proved.

$\square$

Corollary 4.1.6 *Let M be a real analytic six-manifold with a half-flat $H^{\varepsilon,\tau}$-structure that is given by a pair of analytic stable forms (ω_0, ρ_0).*

(i) *Then, there exists a unique maximal solution (ω, ρ) of the evolution equations (4.5), (4.6) with initial value (ω_0, ρ_0), which is defined on an open neighbourhood $\Omega \subset \mathbb{R} \times M$ of $\{0\} \times M$. In particular, there is a parallel $G^{\varepsilon,\tau}$-structure on Ω.*

(ii) *Moreover, the evolution is natural in the sense that, given a diffeomorphism f of M, the pullback $(f^*\omega, f^*\rho)$ of the solution with initial value (ω_0, ρ_0) is the solution of the evolution equations for the initial value $(f^*\omega_0, f^*\rho_0)$.*

In particular, if f is an automorphism of the initial structure (ω_0, ρ_0), then, for all $t \in \mathbb{R}$, f is an automorphism of the solution $(\omega(t), \rho(t))$ defined on the (possibly empty) open set $U_t = \{p \in M \mid (t, p) \in \Omega$ and $(t, f(p)) \in \Omega\}$.

(iii) Furthermore, assume that M is compact or a homogeneous space $M = G/K$ such that the $H^{\varepsilon, \tau}$-structure is G-invariant. Then there is a unique maximal interval $I \ni 0$ and a unique solution (ω, ρ) of the evolution equations (4.5), (4.6) with initial value (ω_0, ρ_0) on $I \times M$. In particular, there is a parallel $G^{\varepsilon, \tau}$-structure on $I \times M$.

Proof If the manifold and the initial structure (ω_0, ρ_0) are analytic, there exists a unique maximal solution of the evolution equations on a neighbourhood Ω of $M \times \{0\}$ in $M \times \mathbb{R}$ by the Cauchy-Kovalevskaya theorem. The naturality of the solution is an immediate consequence of the uniqueness due to the naturality of the exterior derivative. If M is compact, there is a maximal interval I such that the solution is defined on $M \times I$. The same is true for a homogeneous half-flat structure (ω_0, ρ_0) as it is determined by $(\omega_0, \rho_0)|_p$ for any $p \in M$. $\square$

We remark that, for a homogeneous half-flat structure (ω_0, ρ_0), the evolution equations reduce to a system of ordinary differential equations due to the naturality assertion of the corollary. This simplification will be used in Sect. 4.3.3 to construct metrics with holonomy equal to G_2 and G_2^*.

4.1.1 Remark on Completeness: Geodesically Complete Conformal G_2-Metrics

The $G_2^{(*)}$-metrics arising from the Hitchin flow on a six-manifold N are of the form $(I \times N, dt^2 + g_t)$ with an open interval $I = (a, b)$ and a family of Riemannian metrics g_t depending on $t \in I$ (formula (4.10) in Theorem 4.1.3). As curves of the form $t \mapsto (t, x)$ are geodesics for this metric, they are obviously geodesically incomplete if a or $b \in \mathbb{R}$.

For the *Riemannian* case and *compact* manifolds N, we shall explain how one easily obtains *complete* metrics by a conformal change of the G_2-metric.

Lemma 4.1.7 *Let N be a compact manifold with a family g_r of Riemannian metrics. Then the Riemannian metric on $\mathbb{R} \times N$ defined by $h = dr^2 + g_r$ is geodesically complete.*

Proof Denote by d the distance on $\mathbb{R} \times N$ induced by the Riemannian metric $h = dr^2 + g_r$ and by d_r the distance on N induced by g_r. For a curve γ in $M = \mathbb{R} \times N$ we have that the length of $\gamma(t) = (r(t), x(t))$ satisfies

$$\ell(\gamma) = \int_0^1 \sqrt{\dot{r}(t)^2 + g_{r(t)}(\dot{x}(t), \dot{x}(t))}\, dt \geq \int_0^1 |\dot{r}(t)|\, dt \geq |r(1) - r(0)|.$$

As the distance of two points $p = (r, x)$ and $q = (s, y)$ is defined as the infimum of the lengths of all curves joining them, this inequality implies that

$$d(p, q) \geq |r - s|. \tag{4.13}$$

Note also that a curve $\gamma(t) = ((s - r)t + r, x)$ joining $p = (r, x)$ and $q = (s, x)$ in $\mathbb{R} \times \{x\}$ has length $\ell(\gamma) = |r - s|$ and thus, for such p, q we get that $d(p, q) = |r - s|$. On the other hand, for $p = (r, x)$ and $q = (r, y)$ with the same $\mathbb{R}$-projection r we only get that $d(p, q) \leq d_r(x, y)$.

Since h has Riemannian signature we can use the Hopf-Rinow Theorem and consider a Cauchy sequence $p_n = (r_n, x_n) \in \mathbb{R} \times N$ w.r.t. the distance d. Equation (4.13) then implies that the sequence r_n is a Cauchy sequence in $\mathbb{R}$. Hence, r_n converges to $r \in \mathbb{R}$. Since N is compact, the sequence x_n has a subsequence x_{n_k} converging to $x \in N$. For $p = (r, x)$ and $q_{n_k} := (r, x_{n_k})$ the triangle inequality implies that

$$d(p, p_{n_k}) \leq d(p, q_{n_k}) + d(q_{n_k}, p_{n_k}) \leq d_r(x, x_{n_k}) + d(q_{n_k}, p_{n_k}) = d_r(x, x_{n_k}) + |r - r_{n_k}|.$$

Hence, p_{n_k} converges to p. As p_n was a Cauchy sequence, we have found p as a limit for p_n. By the Theorem of Hopf and Rinow, M is geodesically complete. $\square$

The consequence of the lemma is

Proposition 4.1.8 *Let $(M = I \times N, h = dt^2 + g_t)$ be a Riemannian metric on a product of an open interval I and a compact manifold N. Then (M, h) is globally conformally equivalent to a metric on $\mathbb{R} \times N$ that is geodesically complete. The scaling factor depends only on $t \in I$ and is determined by a diffeomorphism $\varphi : \mathbb{R} \to I$.*

Proof Let $\varphi : \mathbb{R} \to I$ be a diffeomorphism with inverse $r = \varphi^{-1}$. Changing the coordinate t to r, the metric h on $I \times N$ can be written as

$$h = (\varphi'(r)dr)^2 + g_{\varphi(r)} = \varphi'(r)^2 \left(dr^2 + \frac{1}{\varphi'(r)^2} g_{\varphi(r)} \right).$$

Hence, h is globally conformally equivalent to the metric $dr^2 + \frac{1}{\varphi'(r)^2} g_{\varphi(r)}$ on $\mathbb{R} \times N$. By the lemma, this metric is geodesically complete. $\square$

Regarding the solution of the Hitchin flow equations, using Theorem 4.1.3, Corollary 4.1.6, and Proposition 4.1.8 we obtain the following consequence.

Corollary 4.1.9 *Let M be a compact analytic six-manifold with half-flat SU(3)-structure given by analytic stable forms (ρ_0, ω_0). Then there is a complete metric on $\mathbb{R} \times M$ that is globally conformal to the parallel G_2-metric obtained by the Hitchin flow.*

In Example 4.3.15 of Sect. 4.3.3 we will construct explicit examples of this type. Finally, note that due to the Cheeger-Gromoll splitting Theorem, see for example

[18, Theorem 6.79], one cannot expect to obtain by the Hitchin flow irreducible G_2-metrics that are complete without allowing degenerations of g_t.

4.1.2 Nearly Half-Flat Structures and Nearly Parallel $G_2^{(*)}$-Structures

A $G_2^{(*)}$-structure φ on a seven-manifold N is called *nearly parallel* if

$$d\varphi = \mu *_\varphi \varphi \qquad (4.14)$$

for a constant $\mu \in \mathbb{R}^*$. Nearly parallel G_2- and G_2^*-structures are also characterised by the existence of a Killing spinor, refer [58] respectively [81].

By Proposition 2.1.14, a $G_2^{(*)}$-structure on a seven-manifold (N, φ) induces an $H^{\varepsilon,\tau}$-structure (ω, ρ) on an oriented hypersurface in (N, φ). If the $G_2^{(*)}$-structure is nearly parallel, the $H^{\varepsilon,\tau}$-structure satisfies the equation $d\rho = -\varepsilon\mu\hat{\omega}$ due to the formulas (2.23) and (2.27). This observation motivates the following definition (see also Sect. 3.1.2 of Chap. 3).

Definition 4.1.10 An $H^{\varepsilon,\tau}$-structure (ω, ρ) on a six-manifold M is called *nearly half-flat* if

$$d\rho = \frac{\lambda}{2}\omega^2 = \lambda\sigma \qquad (4.15)$$

for some constant $\lambda \in \mathbb{R}^*$.

The notion of a nearly half-flat SU(3)-structure was introduced in [53], where also evolution equations on six-manifolds leading to nearly parallel G_2-structures are considered. For compact manifolds M, it is shown in [114] that a solution which is a nearly half-flat SU(3)-structure for a time $t = t_0$ already defines a nearly parallel G_2-structure. In the following, we extend these evolution equations to all possible signatures and give a simplified proof for the properties of the solutions which also holds for non-compact manifolds.

Proposition 4.1.11 *Let $H^{\varepsilon,\tau}$ be a real form of* SL(3, $\mathbb{C}$), *$G^{\varepsilon,\tau}$ the corresponding real form of $G_2^\mathbb{C}$ and (ρ, ω) a one-parameter family of $H^{\varepsilon,\tau}$-structures on a six-manifold M with a parameter t from an interval I. Then, the three-form*

$$\varphi = \omega \wedge dt + \rho$$

defines a nearly parallel $G^{\varepsilon,\tau}$-structure for the constant $\mu \neq 0$ on $M \times I$ if and only if the $H^{\varepsilon,\tau}$-structure (ρ, ω) is nearly half-flat for the constant $-\varepsilon\mu$ for all $t \in I$ and *satisfies the evolution equation*

$$\dot{\rho} = d\omega - \varepsilon\mu\hat{\rho}. \tag{4.16}$$

Proof The assertion follows directly from the following computation, analogously to the proof of Proposition 4.1.2:

$$\check{d}\varphi = d\omega \wedge dt + dt \wedge \dot{\rho} + d\rho \;=\; (d\omega - \dot{\rho}) \wedge dt + d\rho,$$

$$\mu * \varphi = \varepsilon\mu\,(\hat{\rho} \wedge dt - \sigma).$$

□

The main theorem for the parallel case generalises as follows. Recall (2.3) that for a stable four-form $\sigma = \frac{1}{2}\omega^2 = \hat{\omega}$, the application of the operator $\sigma \mapsto \hat{\sigma}$ yields the stable two-form

$$\widehat{\hat{\omega}} = \hat{\sigma} = \frac{1}{2}\omega.$$

Theorem 4.1.12 *Let (ρ_0, ω_0) be a nearly half-flat $H^{\varepsilon,\tau}$-structure for the constant $\lambda \neq 0$ on a six-manifold M. Let M be oriented such that $\omega_0^3 > 0$. Furthermore, let $\rho \in \Omega^3 M$ be a one-parameter family of stable forms with parameters coming from an interval I such that $\rho(t_0) = \rho_0$ and such that the evolution equation*

$$\dot{\rho} = \frac{2}{\lambda}d(\widehat{d\rho}) + \lambda\,\hat{\rho} \tag{4.17}$$

is satisfied for all $t \in I$. Then $(\rho, \omega = \frac{2}{\lambda}\widehat{d\rho})$ is a family of nearly half-flat $H^{\varepsilon,\tau}$-structures for the constant λ. In particular, the three-form

$$\varphi = \omega \wedge dt + \rho$$

defines a nearly parallel $G^{\varepsilon,\tau}$-structure for the constant $-\varepsilon\lambda$ on $M \times I$.

Proof First of all, we observe that $d\rho$ is stable in a neighbourhood of the stable form $d\rho_0 = \lambda\sigma_0$, since stability is an open condition. Furthermore, the operator $d\rho \mapsto \widehat{d\rho}$ is uniquely defined by the orientation induced from ω_0. Therefore, the evolution equation is locally well-defined and we assume that ρ is a solution on an interval I. The only possible candidate for a nearly half-flat structure for the constant λ is $(\rho, \omega = \frac{2}{\lambda}\widehat{d\rho})$ since only this two-form ω satisfies the nearly half-flat equation $\sigma = \hat{\omega} = \frac{1}{\lambda}d\rho$. Obviously, it holds

$$d\sigma = 0 = d\omega \wedge \omega. \tag{4.18}$$

By Proposition 4.1.11, it only remains to show that this pair of stable forms defines an $H^{\varepsilon,\tau}$-structure, or equivalently, that the compatibility conditions (4.1) and (4.2) are preserved in time. By taking the exterior derivative of the evolution equation, we find

$$\dot{\sigma} = \frac{1}{\lambda} d\dot{\rho} = d\hat{\rho} \qquad (4.19)$$

which is in fact the second evolution equation of the parallel case. Completely analogous to the parallel case, the following computation implies the first compatibility condition:

$$
\begin{aligned}
\frac{\partial}{\partial t}(\sigma_X \wedge \rho) &= \dot{\sigma}_X \wedge \rho + \sigma_X \wedge \dot{\rho} \\
&\overset{(4.17),(4.19)}{=} (d\hat{\rho})_X \wedge \rho + \sigma_X \wedge d\omega + \lambda\,\sigma_X \wedge \hat{\rho} \\
&\overset{(4.12),(4.15)}{=} \hat{\rho} \wedge (d\rho)_X + \omega_X \wedge \omega \wedge d\omega + (d\rho)_X \wedge \hat{\rho} \\
&\overset{(4.18)}{=} 0.
\end{aligned}
$$

The proof of the second compatibility condition in [76] again holds literally since the term $\hat{\rho} \wedge \dot{\rho} = \hat{\rho} \wedge d\omega$ is the same as in the case of the parallel evolution. □

The system (4.17) of second order in ρ can easily be reformulated into a system of first order in (ω, ρ) to which we can apply the Cauchy-Kovalevskaya theorem. Indeed, a solution (ω, ρ) of the system

$$\dot{\rho} = d\omega + \lambda\hat{\rho}, \qquad \dot{\sigma} = d\hat{\rho}, \qquad (4.20)$$

with nearly half-flat initial value $(\omega(t_0), \rho(t_0))$ is nearly half-flat for all t and also satisfies the system (4.17). Conversely, (4.17) implies (4.20) with $\sigma = \hat{\omega} = \frac{1}{\lambda} d\rho$.

Therefore, for an initial nearly half-flat structure which satisfies assumptions analogous to those of Corollary 4.1.6, we obtain existence, uniqueness and naturality of a solution of the system (4.20), or, equivalently, of (4.17).

4.1.3 Cocalibrated $G_2^{(*)}$-Structures and Parallel Spin(7)- and $Spin_0(3, 4)$-Structures

In [76], another evolution equation is introduced which relates cocalibrated G_2-structures on compact seven-manifolds M to parallel Spin(7)-structures. As before, we generalise the evolution equation to non-compact manifolds and indefinite metrics.

As we have already seen in Sect. 2.1.3 of Chap. 2, the stabiliser in $GL(V)$ of a four-form Φ_0 on an eight-dimensional vector space V is Spin(7) or $Spin_0(3, 4)$ if and only if it can be written as in (2.30) for a stable three-form φ on a seven-dimensional subspace with stabiliser G_2- or G_2^*, respectively. Thus, a Spin(7)- or

$\mathrm{Spin}_0(3, 4)$-structure on an eight-manifold M is defined by a four-form $\Phi \in \Omega^4 M$ such that $\Phi_p \in \Lambda^4 T_p^* M$ has this property for all p. By formula (2.31) for the metric g_Φ induced by Φ, an oriented hypersurface in (M, Φ) with spacelike unit normal vector field n with respect to g_Φ carries a natural G_2- or G_2^*-structure, respectively, defined by $\varphi = n \lrcorner \Phi$.

A $\mathrm{Spin}(7)$- or $\mathrm{Spin}_0(3, 4)$-structure Φ is *parallel* if and only if $d\Phi = 0$. We remark that the proof for the Riemannian case given in [105, Lemma 12.4] is not hard to transfer to the indefinite case when considering [23, Proposition 2.5] and using the complexification of the two spin groups.

Due to this fact, the induced $G_2^{(*)}$-structure φ on an oriented hypersurface in an eight-manifold M with parallel $\mathrm{Spin}(7)$- or $\mathrm{Spin}_0(3, 4)$-structure Φ is *cocalibrated*, i.e. it satisfies

$$d *_\varphi \varphi = 0. \tag{4.21}$$

Conversely, a cocalibrated $G_2^{(*)}$-structure can be embedded in an eight-manifold with parallel $\mathrm{Spin}(7)$- or $\mathrm{Spin}_0(3, 4)$-structure as follows.

Theorem 4.1.13 *Let M be a seven-manifold and $\varphi \in \Omega^3 M$ be a one-parameter family of stable three-forms with a parameter t in an interval I satisfying the evolution equation*

$$\frac{\partial}{\partial t}(*_\varphi \varphi) = d\varphi. \tag{4.22}$$

If φ is cocalibrated at $t = t_0 \in I$, then φ defines a family of cocalibrated G_2- or $G_2^{(*)}$-structures for all $t \in I$. Moreover, the four-form

$$\Phi = dt \wedge \varphi + *_\varphi \varphi \tag{4.23}$$

defines a parallel $\mathrm{Spin}(7)$- or $\mathrm{Spin}_0(3, 4)$-structure on $M \times I$, respectively, which induces the metric

$$g_\Phi = g_\varphi + dt^2. \tag{4.24}$$

Proof Since the time derivative of $d * \varphi$ vanishes when inserting the evolution equation, the family stays cocalibrated if it is cocalibrated at an initial value. As before, we denote by $\check{d}$ the exterior differential on $\check{M} := M \times I$ and differentiate the four-form (4.23):

$$\check{d}\Phi = -dt \wedge d\varphi + d(*\varphi) + dt \wedge \frac{\partial}{\partial t}(*\varphi).$$

Obviously, this four-form is closed if and only the evolution equation is satisfied and the family is cocalibrated. The formula for the induced metric corresponds to formula (2.31).   □

As before, the Cauchy-Kovalevskaya theorem guarantees existence and uniqueness of solutions if assumptions analogous to those of Corollary 4.1.6 are satisfied.

Remark 4.1.14 We observe that nearly parallel G_2- and G_2^*-structures are in particular cocalibrated such that analytic nearly half-flat structures in dimension 6 can be embedded in parallel Spin(7)- or $\text{Spin}_0(3, 4)$-structures in dimension 8 by evolving them twice with the help of the Theorems 4.1.12 and 4.1.13.

4.2 Evolution of Nearly ε-Kähler Manifolds

In this section, we consider the evolution of nearly pseudo-Kähler and nearly para-Kähler six-manifolds which can be unified by the notion of a nearly ε-Kähler manifold. The explicit solution of the Hitchin flow yields a simple and unified proof for the correspondence of nearly ε-Kähler manifolds and parallel $G_2^{(*)}$-structures on cones. We complete the picture by considering similarly the evolution of nearly Kähler structures to nearly parallel $G_2^{(*)}$-structures on (hyperbolic) sine cones and the evolution of nearly parallel $G_2^{(*)}$-structures to parallel Spin(7)- and $\text{Spin}_0(3, 4)$-structures on cones. Our presentation in terms of differential forms unifies various results in the literature, which were originally obtained using spinorial methods, and applies to all possible real forms of the relevant groups.

4.2.1 Cones over Nearly ε-Kähler Manifolds

For the readers convenience let us shortly recall the setting. In the language of [5] and [110], an *almost ε-Hermitian manifold* (M^{2m}, g, J) is defined by an almost ε-complex structure J which squares to εid and a pseudo-Riemannian metric g which is ε-Hermitian in the sense that $g(J\cdot, J\cdot) = -\varepsilon g(\cdot, \cdot)$. Consequently, a *nearly ε-Kähler manifold* is defined as an almost ε-Hermitian manifold such that ∇J is skew-symmetric. On a six-manifold M, a nearly ε-Kähler structure (g, J, ω) with $|\nabla J|^2 = 4$ (i.e. of constant type 1 in the terminology of [67]) is equivalent to a normalised $H^{\varepsilon,\tau}$-structure (ω, ρ) which satisfies

$$d\omega = 3\rho, \tag{4.25}$$

$$d\hat{\rho} = 4\hat{\omega}. \tag{4.26}$$

This result is well-known for Riemannian signature [103] and is generalised to arbitrary signature in [110, Theorem 3.14], see Sect. 3.1.2 of Chap. 3. In particular, nearly ε-Kähler structures (ω, ρ) in dimension 6 are half-flat and the structure $(\omega, \hat{\rho})$ is nearly half-flat (for the constant $\lambda = 4$).

Proposition 4.2.1 *Let (M, h_0) be a pseudo-Riemannian six-manifold of signature $(6, 0)$, $(4, 2)$ or $(3, 3)$ and let $(\bar{M} = M \times \mathbb{R}^+, \bar{g}_\varepsilon = h_0 - \varepsilon dt^2)$ be the timelike cone for $\varepsilon = 1$ and the spacelike cone for $\varepsilon = -1$. There is a one-to-one correspondence between nearly ε-Kähler structures (h_0, J) with $|\nabla J|^2 = 4$ on (M, h_0) and parallel G_2- and G_2^*-structures φ on $\bar{M}$ which induce the cone metric $\bar{g}_\varepsilon$.*

Proof This well-known fact is usually proved using Killing spinors, see [12, 70] and [82]. We give a proof relying exclusively on the framework of stable forms and the Hitchin flow. For Riemannian signature, this point of view is also adopted in [32] and [25].

The $H^{\varepsilon, \tau}$-structures inducing the given metric h_0 are the reductions of the bundle of orthonormal frames of (M, h_0) to the respective group $H^{\varepsilon, \tau}$. Given any $H^{\varepsilon, \tau}$-reduction (ω_0, ρ_0) of h_0, we consider for $t \in \mathbb{R}^+$ the one-parameter family

$$\omega = t^2 \omega_0, \quad \rho = t^3 \rho_0, \tag{4.27}$$

which induces the family of metrics $h = t^2 h_0$. By formula (4.10), the metric g_φ on $\bar{M}$ induced by the stable three-form $\varphi = \omega \wedge dt + \rho$ is exactly the cone metric $\bar{g}_\varepsilon$.

It is easily verified that the family (4.27) consists of half-flat structures satisfying the evolution equations if and only if the initial value $(\omega(1), \rho(1)) = (\omega_0, \rho_0)$ satisfies the exterior system (4.25), (4.26). Therefore, the stable three-form φ on the cone $(\bar{M}, \bar{g}_\varepsilon)$ is parallel if and only if the $H^{\varepsilon, \tau}$-reduction (ω_0, ρ_0) of h_0 is a nearly ε-Kähler structure with $|\nabla J|^2 = 4$.

Conversely, let φ be a stable three-form on $\bar{M}$ which induces the cone metric $\bar{g}_\varepsilon$. Since ∂_t is a normal vector field for the hypersurface $M = M \times \{1\}$ satisfying $\bar{g}(\partial_t, \partial_t) = -\varepsilon$, we obtain an $H^{\varepsilon, \tau}$-reduction (ω_0, ρ_0) of h_0 defined by

$$\omega_0 = \partial_t \lrcorner \varphi, \qquad \rho_0 = \varphi_{|TM} \tag{4.28}$$

with the help of Proposition 2.1.14 of Chap. 2. Since the two constructions are inverse to each other, the proposition follows. □

Example 4.2.2 Consider the flat $(\mathbb{R}^{(3,4)} \setminus \{0\}, \langle ., . \rangle)$ which is isometric to the cone $(M^\varepsilon \times \mathbb{R}^+, t^2 h_\varepsilon - \varepsilon dt^2)$ over the pseudo-spheres $M^\varepsilon := \{p \in \mathbb{R}^{(3,4)} \mid \langle p, p \rangle = -\varepsilon\}$, $\varepsilon = \pm 1$, with the standard metrics h_ε of constant sectional curvature $-\varepsilon$ and signature $(2, 4)$ for $\varepsilon = -1$ and $(3, 3)$ for $\varepsilon = 1$. Obviously, a stable three-form φ inducing the flat metric $\langle ., . \rangle$ is parallel if and only if it is constant. Thus, the previous discussion and Proposition 2.1.14 of Chap. 2, in particular formula (2.29), yield a bijection

$$SO(3, 4) / G_2^* \to \{\varepsilon\text{-complex structures } J \text{ on } M^\varepsilon \text{ such that } (h_\varepsilon, J) \text{ is nearly } \varepsilon\text{-Kähler}\}$$

$$\varphi \mapsto J \quad \text{with} \quad J_p(v) = -p \times v, \quad \forall p \in M^\varepsilon$$

where the cross-product $\times$ induced by φ is defined by formula (2.12). In other words, the pseudo-spheres $(M^\varepsilon, h_\varepsilon)$ admit a nearly ε-Kähler structure which is unique up to conjugation by the isometry group $O(3, 4)$ of h_ε. In fact, these ε-complex structures on the pseudo-spheres are already considered in [92] and the nearly para-Kähler property for $\varepsilon = 1$ is for instance shown in [15].

4.2.2 Sine Cones over Nearly ε-Kähler Manifolds

For Riemannian signature, it has been shown in [53] that the evolution of a nearly Kähler $SU(3)$-structure to a nearly parallel G_2-structure induces the Einstein sine cone metric. This result can be extended as follows. We prefer to consider (hyperbolic) cosine cones since they are defined on all of $\mathbb{R}$ in the hyperbolic case.

Proposition 4.2.3 *Let (M, h_0) be a pseudo-Riemannian six-manifold.*

(i) *If h_0 is Riemannian, or has signature $(2, 4)$, respectively, there is a one-to-one correspondence between nearly (pseudo-)Kähler structures (h_0, J) on M with $|\nabla J|^2 = 4$ and nearly parallel G_2-structures, or G_2^*-structures, respectively, for the constant $\mu = -4$ on the spacelike cosine cone*

$$(M \times (-\frac{\pi}{2}, \frac{\pi}{2}), \cos^2(t)h_0 + dt^2).$$

(ii) *If h_0 has signature $(3, 3)$, there is a one-to-one correspondence between nearly para-Kähler structures (h_0, J) on M with $|\nabla J|^2 = 4$ and nearly parallel G_2^*-structures for the constant $\mu = 4$ on the timelike hyperbolic cosine cone*

$$(M \times \mathbb{R}, -\cosh^2(t)h_0 - dt^2).$$

Proof

(i) Starting with any $SU(3)$- or $SU(1, 2)$-reduction (ω_0, ρ_0) of h_0, the one-parameter family

$$\omega = \cos^2(t)\omega_0, \quad \rho = -\cos^3(t)(\sin(t)\rho_0 + \cos(t)\hat\rho_0)$$

with $(\omega(0), \rho(0)) = (\omega_0, -\hat\rho_0)$ defines a stable three-form $\varphi = \omega \wedge dt + \rho$ on $M \times (-\frac{\pi}{2}, \frac{\pi}{2})$. Since $z\Psi_0 = z(\rho_0 + i\hat\rho_0)$ is a $(3, 0)$-form w.r.t. the induced almost complex structures $J_{\text{Re}(z\Psi_0)}$ for all $z \in \mathbb{C}^*$, the structure $J_\rho = J_{\rho_0}$ is constant in t. Thus, the metric g_φ induced by φ is the cosine cone metric. Moreover, it holds $\hat\rho = -\cos^3(t)(\sin(t)\hat\rho_0 - \cos(t)\rho_0)$ due to Corollary 2.1.7 of Chap. 2.

It takes a short calculation to verify that the one-parameter family is nearly half-flat (for the constant $\lambda = -4$) and satisfies the evolution equation (4.16) if and only (ω_0, ρ_0) satisfies the exterior system (4.25), (4.26). Thus, applying Proposition 4.1.11, the three-form $\varphi = \omega \wedge dt + \rho$ defines a nearly parallel

$G^{\varepsilon,\tau}$-structure on $M \times (-\frac{\pi}{2}, \frac{\pi}{2})$ (for the constant $\mu = -4$) if and only if (h_0, J_{ρ_0}) is nearly ε-Kähler with $|\nabla J|^2 = 4$.

The inverse construction is given by (4.28) in analogy to the case of the ordinary cone.

(ii) The proof in the para-complex case is completely analogous if we consider the one-parameter family

$$\omega = \cosh^2(t)\omega_0 , \quad \rho = -\cosh^3(t)(\sinh(t)\rho_0 + \cosh(t)\hat{\rho}_0)$$

which is defined for all $t \in \mathbb{R}$. We note the following subtleties regarding signs. By Proposition 2.1.4 of Chap. 2, we know that the mapping $\rho \mapsto \hat{\rho}$ is homogeneous of degree 1, but not linear. Indeed, by applying Corollary 2.1.7 of Chap. 2, we find

$$\overline{\sinh(t)\rho_0 + \cosh(t)\hat{\rho}_0} = -\sinh(t)\hat{\rho}_0 - \cosh(t)\rho_0.$$

Using this formula, one can check that $J_\rho = J_{\hat{\rho}_0} = -J_{\rho_0}$ is constant in t such that the metric induced by (ω, ρ) is in fact $h = -\cosh^2(t)h_0$. $\qquad \square$

The fact that the (hyperbolic) cosine cone over a six-manifold carrying a Killing spinor carries again a Killing spinor was proven in [81]. By relating spinors to differential forms, these results also imply the existence of a nearly parallel $G_2^{(*)}$-structures on the (hyperbolic) cosine cone over a nearly ε-Kähler manifold.

Example 4.2.4 The (hyperbolic) cosine cone of the pseudo-spheres $(M^\varepsilon, h_\varepsilon)$ of Example 4.2.2 has constant sectional curvature 1, for instance due to [8, Corollary 2.3], and is thus (locally) isometric to the pseudo-sphere $S^{3,4} = \{p \in \mathbb{R}^{(4,4)} \mid \langle p, p \rangle = 1\} = \text{Spin}_0(3,4)/G_2^*$.

4.2.3 Cones over Nearly Parallel $G_2^{(*)}$-Structures

By Lemma 9 in [12], there is a one-to-one correspondence on a Riemannian seven-manifold (M, g_0) between nearly parallel G_2-structures and parallel Spin(7)-structures on the Riemannian cone. In order to illustrate the evolution equations for nearly parallel G_2^*-structures, we extend this result to the indefinite case by applying Theorem 4.1.13. This is possible since nearly parallel G_2^*-structures are in particular cocalibrated. Again, the fact that the cone over a nearly parallel G_2^*-manifold admits a parallel spinor can be derived from the connection to Killing spinors as observed in [81].

Proposition 4.2.5 *Let (M, g_0) be a pseudo-Riemannian seven-manifold of signature $(3, 4)$. There is a one-to-one correspondence between nearly parallel*

G_2^*-*structures for the constant* 4 *which induce the given metric* g_0 *and parallel* $\mathrm{Spin}_0(3, 4)$-*structures on* $M \times \mathbb{R}^+$ *inducing the cone metric* $\bar{g} = t^2 g_0 + dt^2$.

Proof Let φ_0 be any cocalibrated G_2^*-structure on M inducing the metric g_0. The one-parameter family of three-forms defined by $\varphi = t^3 \varphi_0$ for $t \in \mathbb{R}^+$ induces the family of metrics $g = t^2 g_0$ such that the Hodge duals are $*_\varphi \varphi = t^4 *_{\varphi_0} \varphi_0$. By (4.24), the $\mathrm{Spin}_0(3, 4)$-structure $\Psi = dt \wedge \varphi + *_\varphi \varphi$ on $M \times \mathbb{R}^+$ induces the cone metric $\bar{g}$. Conversely, given a $\mathrm{Spin}_0(3, 4)$-structure Ψ on the cone $(M \times \mathbb{R}^+, \bar{g})$, we have the cocalibrated G_2^*-structure $\varphi_0 = \partial_t \lrcorner \Psi$ on M, which also induces the given metric g_0. Since the evolution equation (4.22) is satisfied if and only if the initial value φ_0 is nearly parallel for the constant 4 and since the two constructions are inverse to each other, the assertion follows from Theorem 4.1.13.   □

Example 4.2.6 We consider again the easiest example, i.e. the flat $\mathbb{R}^{(4,4)} \setminus \{0\}$ which is isometric to the cone over the pseudo-sphere $S^{3,4}$. Analogous to Example 4.2.2, the proposition just proved yields a proof of the fact that the nearly parallel G_2^*-structures for the constant 4 on $S^{3,4}$ are parametrised by $\mathrm{SO}(4, 4)/\mathrm{Spin}_0(3, 4)$, i.e. by the four homogeneous spaces (2.32). In particular, these structures are conjugated by the isometry group $O(4, 4)$ of $S^{3,4}$.

Summarising the application of the three Propositions 4.2.1, 4.2.3 and 4.2.5 to pseudo-spheres, we find a mutual one-to-one correspondence between

(1) nearly pseudo-Kähler structures with $|\nabla J|^2 \neq 0$ on $(S^{2,4}, g_{can})$,
(2) nearly para-Kähler structures with $|\nabla J|^2 \neq 0$ on $(S^{3,3}, g_{can})$,
(3) parallel G_2^*-structures on $(\mathbb{R}^{(3,4)}, g_{can})$,
(4) nearly parallel G_2^*-structures on the spacelike cosine cone over $(S^{2,4}, g_{can})$,
(5) nearly parallel G_2^*-structures on the timelike hyperbolic cosine cone over $(S^{3,3}, g_{can})$,
(6) nearly parallel G_2^*-structures on $(S^{3,4}, g_{can})$ and
(7) parallel $\mathrm{Spin}_0(3, 4)$-structures on $(\mathbb{R}^{(4,4)}, g_{can})$.

This geometric correspondence is reflected in the algebraic fact that the four homogeneous spaces (2.32) are isomorphic.

4.3 The Evolution Equations on Nilmanifolds $\Gamma \setminus H_3 \times H_3$

Let H_3 be the three-dimensional real Heisenberg group with Lie algebra $\mathfrak{h}_3$. In this section, we will develop a method to explicitly determine the parallel $G_2^{(*)}$-structure induced by an arbitrary invariant half-flat structure on a nilmanifold $\Gamma \setminus H_3 \times H_3$ without integrating. In particular, this method is applied to construct three explicit large families of metrics with holonomy equal to G_2 or G_2^*, respectively.

4.3.1 Evolution of Invariant Half-Flat Structures on Nilmanifolds

Left-invariant half-flat structures (ω_0, ρ_0) on a Lie group G are in one-to-one correspondence with normalised pairs (ω, ρ) of compatible stable forms on the Lie algebra $\mathfrak{g}$ of G which satisfy $d\rho = 0$ and $d\omega^2 = 0$. To shorten the notation, we will speak of a *half-flat structure on a Lie algebra*.

Given as initial value a half-flat structure on a Lie algebra, the evolution equations

$$\dot{\rho} = d\omega , \qquad \dot{\sigma} = d\hat{\rho} , \qquad (4.29)$$

reduce to a system of ordinary differential equations and a unique solution exists on a maximal interval I. Due to the structure of the equation, the solution differs from the initial values by adding exact forms to σ_0 and ρ_0. In other words, an initial value (σ_0, ρ_0) evolves within the product $[\sigma_0] \times [\rho_0]$ of their respective Lie algebra cohomology classes.

Every nilpotent Lie group N with rational structure constants admits a co-compact lattice Γ and the resulting compact quotients $\Gamma \setminus N$ are called nilmanifolds. Recall that a geometric structure on a nilmanifold $\Gamma \setminus N$ is called *invariant* if is induced by a left-invariant geometric structure on N.

Explicit solutions of the Hitchin flow equations on several nilpotent Lie algebras can be found for instance in [31] and [2]. In both cases, a metric with holonomy contained in G_2 has been constructed before by a different method and this information is used to obtain the solution. For a symplectic half-flat initial value, another explicit solution on one of these Lie algebras is given in [37]. In all cases, the solution depends only on one variable.

At least for four nilpotent Lie algebras including $\mathfrak{h}_3 \oplus \mathfrak{h}_3$, a reason for the simple structure of the solutions has been observed in [2]. Indeed, the following lemma shows that the evolution of σ takes place in a one-dimensional space. As usual, we define a nilpotent Lie algebra by giving the image of a basis of one-forms under the exterior derivative, see for instance [105]. The same reference also contains a list of all six-dimensional nilpotent Lie algebras.

Lemma 4.3.1 *Let ρ be a closed stable three-form with dual three-form $\hat{\rho}$ on a six-dimensional nilpotent Lie algebra $\mathfrak{g}$.*

(i) If $\mathfrak{g}$ is one of the three Lie algebras

$$(0,0,0,0,e^{12},e^{34}) , \quad (0,0,0,0,e^{13}+e^{42},e^{14}+e^{23}), \quad (0,0,0,0,e^{12},e^{14}+e^{23}),$$

then $d\hat{\rho} \in \Lambda^4 U$ for the four-dimensional kernel U of $d : \Lambda^1 \mathfrak{g}^* \to \Lambda^2 \mathfrak{g}^*$.

(ii) If $\mathfrak{g}$ is the Lie algebra

$$(0,0,0,0,0,e^{12} + e^{34}),$$

then $d\hat{\rho} \in \Lambda^4 U$ for the four-dimensional subspace $U = \mathrm{span}\{e^1, e^2, e^3, e^4\}$ of $\ker d$.

Remark 4.3.2 The assertion of the lemma is not true for the remaining six-dimensional nilpotent Lie algebras with $b_1 = \dim(\ker d) = 4$ or $b_1 = 5$. In each case, we have constructed a closed stable ρ such that $d\hat{\rho}$ is not contained in $\Lambda^4(\ker d)$.

In fact, this lemma can also be viewed as a corollary of the following lemma which we will prove first.

Lemma 4.3.3 Let ρ be a closed stable three-form on one of the four Lie algebras of Lemma 4.3.1 and let U be the four-dimensional subspace of $\ker d$ defined there. In all four cases, the space U is J_ρ-invariant where J_ρ denotes the almost (para-)complex structure induced by ρ.

Proof For $\lambda(\rho) < 0$, the assertion is similar to that of [2, Lemma 2]. However, since the only proof seems to be given for the Iwasawa algebra for integrable J in [83, Theorem 1.1], we give a complete proof.

Let $\mathfrak{g}$ be one of the three Lie algebras given in part (i) of Lemma 4.3.1 and $U = \ker d$. Obviously, the two-dimensional image of d lies within $\Lambda^2 U$ in all three cases. By $J = J_\rho$ we denote the almost (para-)complex structure associated to the closed stable three-form ρ. As before, we denote by $\varepsilon \in \{\pm 1\}$ the sign of $\lambda(\rho)$ such that $J_\rho = \varepsilon\mathrm{id}$. Let the symbol i_ε be defined by the property $i_\varepsilon^2 = \varepsilon$ such that the para-complex numbers and the complex numbers can be unified by $\mathbb{C}_\varepsilon = \mathbb{R}[i_\varepsilon]$. Thus, a $(1,0)$-form can be defined for both values of ε as an eigenform of J_ρ in $\Lambda^1 \mathfrak{g}^* \otimes \mathbb{C}_\varepsilon$ for the eigenvalue i_ε.

We define the J-invariant subspace $W := U \cap J^* U$ of $\mathfrak{g}$ such that $2 \le \dim W \le 4$. In fact, $\dim W = 4$ is equivalent to the assertion. The other two cases are not possible, which can be seen as follows. To begin with, assume that W is two-dimensional. When choosing a complement W' of W in U, we have by definition of W that

$$V = W \oplus W' \oplus J^* W'.$$

We observe that, for $\varepsilon = 1$, the ± 1-eigenspaces of J restricted to $W' \oplus J^* W'$ are both two-dimensional. Therefore, we can choose for both values of ε a basis $\{e^1, e^2, e^3, e^4 = J^* e^1, e^5 = J^* e^2, e^6 = J^* e^3\}$ of V such that e^1, e^2, e^3 and e^4 are closed and $de^5, de^6 \in \Lambda^2 U$. Since $\rho + i_\varepsilon J_\rho^* \rho$ is a $(3,0)$-form in both cases, it is possible to change the basis vectors e^1, e^4 within $W \subset \ker d$ such that

$$\rho + i_\varepsilon J_\rho^* \rho = (e^1 + i_\varepsilon e^4) \wedge (e^2 + i_\varepsilon e^5) \wedge (e^3 + i_\varepsilon e^6)$$

and thus

$$\rho = e^{123} + \varepsilon e^{156} - \varepsilon e^{246} + \varepsilon e^{345}.$$

By construction of the basis, we have that

$$0 = d\rho = -\varepsilon e^1 \wedge de^5 \wedge e^6 + \varepsilon e^1 \wedge e^5 \wedge de^6 + \alpha$$

with $\alpha \in \Lambda^4 U$. As the first two summands are linearly independent and not in $\Lambda^4 U$, we conclude that both $e^1 \wedge de^5$ and $e^1 \wedge de^6$ vanish. Thus, the closed one-form e^1 has the property that the wedge product of e^1 with any exact two-form vanishes. However, an inspection of the standard basis of each of the three Lie algebras in question reveals that such a one-form does not exist on these Lie algebras and we have a contradiction to dim $W = 2$.

Since a J-invariant space cannot be three-dimensional for $\varepsilon = -1$, the proof is finished for this case. However, if $\varepsilon = 1$, the case dim $W = 3$ cannot be excluded that easy. Assuming that it is in fact dim $W = 3$, we choose again a complement W' of W in U and find a decomposition

$$V = W \oplus W' \oplus J^* W' \oplus W''$$

with $J^* W'' = W''$. Without restricting generality, we can assume that J acts trivially on W''. Then, we find a basis for V such that the $+1$-eigenspace of J is spanned by $\{e^1, e^4 + e^5, e^6\}$ and the -1-eigenspace by $\{e^2, e^3, e^4 - e^5\}$, where e^1, e^2, e^3 and e^4 are closed and $e^5 = J^* e^4$. Since the given closed three-form ρ generates this J, it has to be of the form

$$\rho = ae^1 \wedge (e^4 + e^5) \wedge e^6 + be^{23} \wedge (e^4 - e^5)$$

for two real constants a, b. The vanishing exterior derivative

$$d\rho = ae^1 \wedge d(e^{56}) \qquad \text{mod} \quad \Lambda^4 U$$

leads to the same contradiction as in the first case and part (i) is shown.

In fact, the same arguments apply to the Lie algebra of part (ii). The four-dimensional space $U \subset \ker d$ spanned by $\{e^1, \ldots, e^4\}$ also satisfies im $d \subset \Lambda^2 U$. Going through the above arguments, the only difference is that e^5 or e^6 may be closed. However, at least one of them is not closed and its image under d generates the exact two-forms. Again, there is no one-form $\beta \in U$ such that $\beta \wedge \gamma = 0$ for all exact two-forms γ and the arguments given in part (i) lead to contradictions for both dim $W = 2$ and dim $W = 3$. $\qquad\square$

Proof of Lemma 4.3.1 Let ρ be a closed stable three-form on one of the four nilpotent Lie algebras and $U \subset \ker d$ as defined in the lemma. For both values of ε, we can apply Lemma 4.3.3 and choose two linearly independent closed $(1,0)$-forms E^1 and E^2 within the J_ρ-invariant space $U \otimes \mathbb{C}_\varepsilon$. Considering that $\rho + i_\varepsilon \hat\rho$ is a $(3,0)$-form for both values of ε, there is a third $(1,0)$-form E^3 such that $\rho + i_\varepsilon \hat\rho = E^{123}$.

Since $d\rho = 0$ and im $d \subset \Lambda^2 U$, it follows that the exterior derivative

$$d\hat{\rho} = \varepsilon i_\varepsilon d(E^{123}) = \varepsilon i_\varepsilon E^{12} \wedge dE^3$$

is an element of $\Lambda^4 U$.   □

4.3.2 Left-Invariant Half-Flat Structures on $H_3 \times H_3$

From now on, we focus on the Lie algebra $\mathfrak{g} = \mathfrak{h}_3 \oplus \mathfrak{h}_3$. Apart from describing all half-flat structures on this Lie algebra, i.e. all initial values for the evolution equations, we give various explicit examples and prove a strong rigidity result concerning the induced metric.

Obviously, pairs of compatible stable forms on a Lie algebra which are isomorphic by a Lie algebra automorphism induce equivalent $H^{\varepsilon,\tau}$-structures on the corresponding simply connected Lie group. Thus, we derive, to begin with, a normal form modulo Lie algebra automorphisms for stable two-forms $\omega \in \Lambda^2 \mathfrak{g}^*$ which satisfy $d\omega^2 = 0$.

A basis $\{e_1, e_2, e_3, f_1, f_2, f_3\}$ for $\mathfrak{h}_3 \oplus \mathfrak{h}_3$ such that the only non-vanishing Lie brackets are given by

$$de^3 = e^{12}, \qquad df^3 = f^{12},$$

will be called a *standard basis*. The connected component of the automorphism group of the Lie algebra $\mathfrak{h}_3 \oplus \mathfrak{h}_3$ in the standard basis is

$$\mathrm{Aut}_0(\mathfrak{h}_3 \oplus h_3) = \left\{ \begin{pmatrix} A & 0 & 0 & 0 \\ a^t \det(A) & c^t & 0 \\ 0 & 0 & B & 0 \\ d^t & 0 & b^t \det(B) \end{pmatrix}, A, B \in \mathrm{GL}(2, \mathbb{R}), a, b, c, d \in \mathbb{R}^2 \right\}.$$

$$\text{(4.30)}$$

We denote by $\mathfrak{g}_i$, $i = 1, 2$, the two summands, by $\mathfrak{z}_i$ their centres and by $\mathfrak{z}$ the centre of $\mathfrak{g}$. The annihilator of the centre is $\mathfrak{z}^0 = \ker d$ and similarly for the summands by restricting d. We have the decompositions

$$\mathfrak{g}^* \cong \mathfrak{z}_1^0 \oplus \mathfrak{z}_2^0 \oplus \frac{\mathfrak{g}_1^*}{\mathfrak{z}_1^0} \oplus \frac{\mathfrak{g}_2^*}{\mathfrak{z}_2^0},$$

$$\Lambda^2 \mathfrak{g}^* \cong \Lambda^2(\mathfrak{z}^0) \oplus \underbrace{\left(\frac{\mathfrak{g}_1^*}{\mathfrak{z}_1^0} \wedge \frac{\mathfrak{g}_2^*}{\mathfrak{z}_2^0}\right)}_{\ell_1} \oplus \underbrace{\left(\mathfrak{z}_1^0 \wedge \frac{\mathfrak{g}_2^*}{\mathfrak{z}_2^0}\right) \oplus \left(\mathfrak{z}_2^0 \wedge \frac{\mathfrak{g}_1^*}{\mathfrak{z}_1^0}\right)}_{\ell_2 \qquad\qquad \ell_3} \oplus \underbrace{\left(\mathfrak{z}_1^0 \wedge \frac{\mathfrak{g}_1^*}{\mathfrak{z}_1^0}\right) \oplus \left(\mathfrak{z}_2^0 \wedge \frac{\mathfrak{g}_2^*}{\mathfrak{z}_2^0}\right)}_{\ell_4}.$$

By $\omega^{\mathfrak{k}_i}$ we denote the projection of a two-form ω onto one of the spaces $\mathfrak{k}_i$, $i = 1, 2, 3, 4$, defined as indicated in the decomposition. We observe that $\mathfrak{k}_1 = \Lambda^2(\frac{\mathfrak{g}^*}{\mathfrak{z}^0})$ and $\omega^{\mathfrak{k}_1} = 0$ if and only if $\omega(\mathfrak{z}, \mathfrak{z}) = 0$.

Lemma 4.3.4 *Consider the action of* $\mathrm{Aut}(\mathfrak{h}_3 \oplus \mathfrak{h}_3)$ *on the set of non-degenerate two-forms* ω *on* $\mathfrak{g}$ *with* $d\omega^2 = 0$. *The orbits modulo rescaling are represented in a standard basis by the following two-forms:*

$$
\begin{aligned}
\omega_1 &= e^1 f^1 + e^2 f^2 + e^3 f^3, & &\text{if } \omega^{\mathfrak{k}_1} \neq 0, \\
\omega_2 &= e^2 f^2 + e^{13} + f^{13}, & &\text{if } d\omega = 0 \iff \omega^{\mathfrak{k}_1} = 0, \omega^{\mathfrak{k}_2} = 0, \omega^{\mathfrak{k}_3} = 0, \\
\omega_3 &= e^1 f^3 + e^2 f^2 + e^3 f^1, & &\text{if } \omega^{\mathfrak{k}_1} = 0, \omega^{\mathfrak{k}_2} \neq 0, \omega^{\mathfrak{k}_3} \neq 0, \omega^{\mathfrak{k}_4} = 0, \\
\omega_4 &= e^1 f^3 + e^2 f^2 + e^3 f^1 + e^{13} + \beta f^{13}, & &\text{if } \omega^{\mathfrak{k}_1} = 0, \omega^{\mathfrak{k}_2} \neq 0, \omega^{\mathfrak{k}_3} \neq 0, \omega^{\mathfrak{k}_4} \neq 0, \\
\omega_5 &= e^1 f^3 + e^2 f^2 + e^{13} + f^{13} & &\text{otherwise,}
\end{aligned}
$$

where $\beta \in \mathbb{R}$ *and* $\beta \neq -1$.

Proof Let

$$
\omega = \sum \alpha_i e^{(i+1)(i+2)} + \sum \beta_i f^{(i+1)(i+2)} + \sum \gamma_{i,j} e^i f^j
$$

be an arbitrary non-degenerate two-form expressed in a standard basis. We will give in each case explicitly a change of standard basis by an automorphism of the form (4.30) with the notation

$$
A = \begin{pmatrix} a_1 & a_2 \\ a_3 & a_4 \end{pmatrix}, \; a^t = (a_5, a_6), \; B = \begin{pmatrix} b_1 & b_2 \\ b_3 & b_4 \end{pmatrix}, \; b^t = (b_5, b_6), \; c^t = (c_1, c_2), \; d^t = (d_1, d_2).
$$

First of all, if $\omega^{\mathfrak{k}_1} \neq 0$, the term $\gamma_{3,3} e^3 f^3$ is different from zero and we rescale such that $\gamma_{3,3} = 1$. Then, the application of the change of basis

$$
\begin{aligned}
&a_1 = 1, \; a_2 = 0, \; a_3 = 0, \; a_4 = 1, \; a_5 = -\gamma_{1,3}, \; a_6 = -\gamma_{2,3}, \\
&b_1 = \gamma_{2,2} - \gamma_{2,3}\gamma_{3,2} - \alpha_1\beta_1, \; b_2 = -\gamma_{1,2} - \beta_1\alpha_2 + \gamma_{3,2}\gamma_{1,3}, \\
&b_3 = -\gamma_{2,1} + \gamma_{3,1}\gamma_{2,3} - \beta_2\alpha_1, \\
&b_4 = \gamma_{1,1} - \alpha_2\beta_2 - \gamma_{1,3}\gamma_{3,1}, \; b_5 = -\gamma_{3,1}\gamma_{2,2} + \gamma_{3,1}\alpha_1\beta_1 + \gamma_{3,2}\gamma_{2,1} + \gamma_{3,2}\beta_2\alpha_1, \\
&b_6 = \gamma_{3,1}\gamma_{1,2} + \gamma_{3,1}\beta_1\alpha_2 - \gamma_{3,2}\gamma_{1,1} + \gamma_{3,2}\alpha_2\beta_2, \\
&c_1 = \beta_2\gamma_{2,2} - \beta_2\gamma_{2,3}\gamma_{3,2} + \beta_1\gamma_{2,1} - \beta_1\gamma_{3,1}\gamma_{2,3}, \; d_1 = -\alpha_2, \\
&c_2 = -\beta_2\gamma_{1,2} + \beta_2\gamma_{3,2}\gamma_{1,3} - \beta_1\gamma_{1,1} + \beta_1\gamma_{1,3}\gamma_{3,1}, \; d_2 = \alpha_1,
\end{aligned}
$$

transforms ω into $\tilde{\omega} = \tilde{\gamma}_{1,1}(e^1 f^1 + e^2 f^2 + e^3 f^3) + \tilde{\alpha}_3 e^{12} + \tilde{\beta}_3 f^{12}$, $\tilde{\gamma}_{1,1} \neq 0$. This two-form satisfies $d\tilde{\omega}^2 = 0$ if and only if $\tilde{\alpha}_3 = 0, \tilde{\beta}_3 = 0$ and the normal form ω_1 is achieved by rescaling.

Secondly, the vanishing of $d\omega$ corresponds to $\omega^{t_1} = 0$, $\omega^{t_2} = 0$, $\omega^{t_3} = 0$ or $\gamma_{3,3} = \gamma_{1,3} = \gamma_{2,3} = \gamma_{3,1} = \gamma_{3,2} = 0$ in a standard basis. By non-degeneracy, at least one of α_1 and α_2 is not zero and we can always achieve $\alpha_1 = 0, \alpha_2 \neq 0$. Indeed, if $\alpha_1 \neq 0$, we apply the transformation (4.30) with $a_1 = 1, a_2 = 1, a_4 = \frac{\alpha_2}{\alpha_1}$, $B = \mathbb{1}$ and all remaining entries zero. With an analogous argument, we can assume that $\beta_1 = 0, \beta_2 \neq 0$. Since $\gamma_{2,2} \neq 0$ by non-degeneracy, we can rescale ω such that $\gamma_{2,2} = 1$. Now, the transformation of the form (4.30) given by

$$a_1 = 1, \ a_2 = 0, \ a_3 = 0, \ a_4 = -\beta_2, \ b_1 = 1, \ b_2 = 0, \ b_3 = 0, \ b_4 = -\alpha_2, \ a_5 = 0,$$

$$a_6 = -\frac{\alpha_3 \beta_2}{\alpha_2}, \ b_5 = 0, \ b_6 = -\frac{\alpha_2 \beta_3}{\beta_2}, \ c_1 = \frac{\gamma_{1,1}}{\alpha_2}, \ c_2 = -\gamma_{1,2}, \ d_1 = 0, \ d_2 = \gamma_{2,1},$$

maps ω to a multiple of the normal form ω_2.

Thirdly, we assume that ω is non-degenerate with $\omega^{t_1} = 0$, i.e. $\gamma_{3,3} = 0$ and both $\omega^{t_2} \neq 0$, i.e. $\gamma_{1,3}$ or $\gamma_{2,3} \neq 0$, and $\omega^{t_3} \neq 0$, i.e. $\gamma_{3,1}$ or $\gamma_{3,2} \neq 0$. Similar as before, we can achieve $\gamma_{2,3} = 0$, $\gamma_{1,3} \neq 0$ by applying, if $\gamma_{2,3} \neq 0$, the transformation (4.30) with $a_1 = 1$, $a_2 = 1$, $a_4 = -\frac{\gamma_{1,3}}{\gamma_{2,3}}$, $B = \mathbb{1}$ and all remaining entries zero. Analogously, we can assume $\gamma_{3,2} = 0$, $\gamma_{3,1} \neq 0$ and rescaling yields $\gamma_{2,2} = 1$, which is non-zero by non-degeneracy. After this simplification, the condition $d\omega^2 = 0$ implies that $\alpha_1 = \beta_1 = 0$ and the transformation

$$a_1 = 1, \ a_2 = 0, \ a_3 = \frac{\alpha_2 \beta_3 - \gamma_{3,1}\gamma_{1,2}}{\gamma_{3,1}}, \ a_4 = \gamma_{1,3}, \ a_5 = 0, \ a_6 = 0, \ b_1 = 1, \ b_2 = 0,$$

$$b_3 = \frac{\beta_2 \alpha_3 - \gamma_{1,3}\gamma_{2,1}}{\gamma_{1,3}}, \ b_4 = \gamma_{3,1}, \ b_5 = \frac{\gamma_{1,2}\gamma_{1,3}\gamma_{2,1}\gamma_{3,1} - \gamma_{1,1}\gamma_{1,3}\gamma_{3,1} - \alpha_2\alpha_3\beta_2\beta_3}{\gamma_{1,3}^2 \gamma_{3,1}},$$

$$b_6 = 0,$$

$$c_1 = 0, \ c_2 = \beta_3, \ d_1 = 0, \ d_2 = -\alpha_3,$$

maps ω to $\tilde{\omega} = e^1 f^3 + e^2 f^2 + e^3 f^1 + \tilde{\alpha}_2 e^{31} + \tilde{\beta}_2 f^{31}$. The condition $\omega^{t_4} = 0$ corresponds to $\tilde{\alpha}_2 = 0, \tilde{\beta}_2 = 0$, i.e. normal form ω_3. If $\omega^{t_4} \neq 0$, we can achieve $\tilde{\alpha}_2 \neq 0$ by possibly changing the summands. Now, the transformation

$$a_1 = 1, \ a_2 = 0, \ a_3 = 0, \ a_4 = -\frac{1}{\tilde{\alpha}_2}, \ a_5 = 0, \ a_6 = 0, \ c_1 = 0, \ c_2 = 0,$$

$$b_1 = -\frac{1}{\tilde{\alpha}_2}, \ b_2 = 0, \ b_3 = 0, \ b_4 = -\frac{1}{\tilde{\alpha}_2}, \ b_5 = 0, \ b_6 = 0, \ d_1 = 0, \ d_2 = 0,$$

maps $\tilde{\omega}$ to the fourth normal form ω_4.

The cases that remain are $\omega^{t_1} = 0$ and either $\omega^{t_2} \neq 0, \omega^{t_3} = 0$ or $\omega^{t_3} = 0, \omega^{t_2} \neq 0$. After changing the summands if necessary, we can assume $\omega^{t_3} = 0$ and

$\omega^{t_2} \neq 0$, i.e. $\gamma_{3,1} = \gamma_{3,2} = \gamma_{3,3} = 0$ and at least one of $\gamma_{1,3}$ or $\gamma_{2,3}$ non-zero. As before, we can achieve $\gamma_{2,3} = 0$ by the transformation $a_1 = 1, a_2 = 1, a_4 = -\frac{\gamma_{1,3}}{\gamma_{2,3}}$. Evaluating $d\omega^2 = 0$ yields $\alpha_1 = 0$. Now, non-degeneracy enforces that $\beta_1 \neq 0$ or $\beta_2 \neq 0$, and after another similar transformation $\beta_1 = 0$. Finally, the simplified ω is non-degenerate if and only if $\gamma_{2,2}\alpha_2\beta_2 \neq 0$ and, after rescaling such that $\gamma_{2,2} = 1$, the transformation

$$a_1 = 1, \quad a_2 = 0, \quad a_3 = 0, \quad a_4 = -\frac{\gamma_{1,3}^2}{\beta_2}, \quad a_5 = 0, \quad a_6 = \frac{\gamma_{1,3}^2(\gamma_{1,3}\gamma_{2,1} - \alpha_3\beta_2)}{\alpha_2\beta_2^2},$$

$$b_1 = -\frac{\gamma_{1,3}}{\beta_2}, \quad b_2 = 0, \quad b_3 = 0, \quad b_4 = -\alpha_2, \quad b_5 = 0, \quad b_6 = -\frac{\beta_3\alpha_2}{\beta_2},$$

$$c_1 = 0, \quad c_2 = -\frac{\gamma_{1,2}\beta_2 + \gamma_{1,3}\beta_3}{\beta_2}, \quad d_1 = -\frac{\gamma_{1,1}}{\beta_2}, \quad d_2 = \frac{\gamma_{1,3}^2\gamma_{2,1}}{\beta_2^2},$$

maps ω to a multiple of the fifth normal form ω_5. $\square$

Using this lemma, it is possible to describe all half-flat structures (ω, ρ) on $\mathfrak{h}_3 \oplus \mathfrak{h}_3$ as follows. In a fixed standard basis such that ω is in one of the normal forms, the equations $d\rho = 0$ and $\omega \wedge \rho = 0$ are linear in the coefficients of an arbitrary three-form ρ. Thus, it is straightforward to write down all compatible closed three-forms for each normal form which depend on nine parameters in each case. The stable forms in this nine-dimensional space are parametrised by the complement of the zero-set of the polynomial $\lambda(\rho)$ of order four. One parameter is eliminated when we require a stable ρ to be normalised in the sense of (2.18). We remark that the computation of the induced tensors J_ρ, $\hat{\rho}$ and $g_{(\omega,\rho)}$ may require computer support, in particular, the signature of the metric is not obvious. However, stability is an open condition: If a single half-flat structure (ω_0, ρ_0) is explicitly given such that ω_0 is one of the normal forms, then the eight-parameter family of normalised compatible closed forms defines a deformation of the given half-flat structure (ω_0, ρ_0) in some neighbourhood of (ω_0, ρ_0).

For instance, the closed three-forms which are compatible with the first normal form

$$\omega = e^1 f^1 + e^2 f^2 + e^3 f^3 \tag{4.31}$$

in a standard basis can be parametrised as follows:

$$\begin{aligned}
\rho = \rho(a_1, \ldots, a_9) &= a_1\, e^{123} + a_2 f^{123} + a_3\, e^1 f^{23} + a_4\, e^2 f^{13} \\
&+ a_5\, e^{23} f^1 + a_6\, e^{13} f^2 + a_7\, (e^2 f^{23} - e^1 f^{13}) + a_8\, (e^{12} f^3 - e^3 f^{12}) \\
&+ a_9\, (e^{23} f^2 - e^{13} f^1).
\end{aligned} \tag{4.32}$$

The quartic invariant $\lambda(\rho)$ depending on the nine parameters is

$$
\begin{aligned}
\lambda(\rho) = (&2a_6a_4a_8^2 + 2a_1a_2a_8^2 + 2a_8^2a_3a_5 - 4a_5a_7^2a_6 - 4a_9^2a_4a_3 - 4a_9^2a_2a_8 + 4a_7^2a_8a_1 \\
&+4a_7a_8^2a_9 + a_1^2a_2^2 + a_6^2a_4^2 + a_3^2a_5^2 + a_8^4 - 2a_6a_4a_3a_5 + 4a_5a_7a_9a_3 + 4a_9a_4a_6a_7 \\
&-4a_5a_2a_6a_8 + 4a_4a_8a_1a_3 - 4a_9a_2a_1a_7 - 2a_1a_2a_6a_4 - 2a_1a_2a_3a_5) \left(e^{123}f^{123}\right)^{\otimes 2}.
\end{aligned}
$$

Example 4.3.5 For each possible signature, we give an explicit normalised half-flat structure with fundamental two-form (4.31). The first and the third example appear in [112]. To begin with, the closed three-form

$$
\rho = \frac{1}{\sqrt{2}}(e^{123} - f^{123} - e^1 f^{23} + e^{23} f^1 - e^2 f^{31} + e^{31} f^2 - e^3 f^{12} + e^{12} f^3) \quad (4.33)
$$

induces a half-flat SU(3)-structure (ω, ρ) such that the standard basis is orthonormal. Similarly, the closed three-form

$$
\rho = \frac{1}{\sqrt{2}}(e^{123} - f^{123} - e^1 f^{23} + e^{23} f^1 + e^2 f^{31} - e^{31} f^2 + e^3 f^{12} - e^{12} f^3) \quad (4.34)
$$

induces a half-flat SU(1, 2)-structure (ω, ρ) such that the standard basis is pseudo-orthonormal with e_1 and e_4 being spacelike. Finally, the closed three-form

$$
\rho = \sqrt{2}\,(e^{123} + f^{123}), \quad (4.35)
$$

induces a half-flat SL(3, $\mathbb{R}$)-structure (ω, ρ) such that the two $\mathfrak{h}_3$-summands are the eigenspaces of the para-complex structure J_ρ, which is integrable since also $d\hat{\rho} = 0$. The induced metric is

$$
g = 2\left(e^1 \cdot e^4 + e^2 \cdot e^5 + e^3 \cdot e^6\right).
$$

In fact, half-flat structures with Riemannian metrics are only possible if ω belongs to the orbit of the first normal form.

Lemma 4.3.6 *Let (ω, ρ) be a half-flat SU(3)-structure on $\mathfrak{h}_3 \oplus \mathfrak{h}_3$. Then it holds $\omega^{\mathfrak{k}_1} \neq 0$. In particular, there is a standard basis such that $\omega = \omega_1 = e^1 f^1 + e^2 f^2 + e^3 f^3$.*

Proof Suppose that (ω, ρ) is a half-flat SU(3)-structure on $\mathfrak{h}_3 \oplus \mathfrak{h}_3$ with $\omega^{\mathfrak{k}_1} = 0$. Thus, we can choose a standard basis such that ω is in one of the normal forms $\omega_2, \ldots, \omega_5$ of Lemma 4.3.4 and ρ belongs to the corresponding nine-parameter family of compatible closed three-forms. We claim that the basis one-form e^1 is isotropic in all four cases which yields a contradiction since the metric of an SU(3)-structure is positive definite. The quickest way to verify the claim is the direct computation of the induced metric, which depends on nine parameters, with the help of a computer. In order to verify the assertion by hand, the following formulas shorten the calculation considerably. For all one-forms α, β and all vectors v, the ε-complex structure J_ρ and the metric g induced by a compatible pair (ω, ρ) of stable forms satisfy

$$\alpha \wedge J_\rho^* \beta \wedge \omega^2 = g(\alpha, \beta) \frac{1}{3} \omega^3,$$

$$J_\rho^* \alpha(v) \phi(\rho) = \alpha \wedge \rho \wedge (v \lrcorner \rho),$$

which is straightforward to verify in the standard basis (2.7), (2.20), cf. also [112, Lemmas 2.1,2.2]. For instance, for the second normal form ω_2, it holds $e^1 \wedge \omega_2^2 = -2e^{12}f^{123}$. Thus, by the first formula, it suffices to show that $J_\rho^* e^1(e_3) = e^1(J_\rho e_3) = 0$ which is in turn satisfied if $e^1 \wedge \rho \wedge (e_3 \lrcorner \rho) = 0$ due to the second formula. A similar simplification applies to the other normal forms and we omit the straightforward calculations.   □

Moreover, the geometry turns out to be very rigid if $\omega^{\mathfrak{e}_1} = 0$. We recall that simply connected para-hyper-Kähler symmetric spaces with abelian holonomy are classified in [7, 45]. In particular, there exists a unique simply connected four-dimensional para-hyper-Kähler symmetric space with one-dimensional holonomy group, which is defined in [7, Section 4]. We denote the underlying pseudo-Riemannian manifold as (N^4, g_{PHK}).

Proposition 4.3.7 *Let (ω, ρ) be a left-invariant half-flat structure with $\omega^{\mathfrak{e}_1} = 0$ on $H_3 \times H_3$ and let g be the pseudo-Riemannian metric induced by (ω, ρ). Then, the pseudo-Riemannian manifold $(H_3 \times H_3, g)$ is either flat or isometric to the product of (N^4, g_{PHK}) and a two-dimensional flat factor. In particular, the metric g is Ricci-flat.*

Proof Due to the assumption $\omega^{\mathfrak{e}_1} = 0$, we can choose a standard basis such that ω is in one of the normal forms $\omega_2, \dots, \omega_5$. In each case separately, we do the following. We write down all compatible closed three-forms ρ depending on nine parameters. With computer support, we calculate the induced metric g. For the curvature considerations, it suffices to work up to a constant such that we can ignore the rescaling by $\lambda(\rho)$ which is different from zero by assumption. Now, we transform the left-invariant co-frame $\{e^1, \dots, f^3\}$ to a coordinate co-frame $\{dx_1, \dots, dy_3\}$ by applying the transformation defined by

$$e^1 = dx_1, \ e^2 = dx_2, \ e^3 = dx_3 + x_1 dx_2, \ f^1 = dy_1, \ f^2 = dy_2, \ f^3 = dy_3 + y_1 dy_2, \tag{4.36}$$

such that the metric is accessible for any of the numerous packages computing curvature. The resulting curvature tensor $R \in \Gamma(\text{End } \Lambda^2 TM)$, $M = H^3 \times H^3$, has in each case only one non-trivial component

$$R(\partial_{x_1} \wedge \partial_{y_1}) = c \, \partial_{x_3} \wedge \partial_{y_3} \tag{4.37}$$

for a constant $c \in \mathbb{R}$ and R is always parallel. Thus, the metric is flat if $c = 0$ and symmetric with one-dimensional holonomy group if $c \neq 0$, for $H_3 \times H_3$ is simply connected and a naturally reductive homogeneous metric is complete.

Furthermore, it turns out that the metric restricted to $TN := span\{\partial_{x_1}, \partial_{x_3}, \partial_{y_1}, \partial_{y_3}\}$ is non-degenerate and of signature $(2, 2)$ for all parameter values. Thus, the manifold splits in a four-dimensional symmetric factor with neutral metric and curvature tensor (4.37) and the two-dimensional orthogonal complement which is flat. Since a simply connected symmetric space is completely determined by its curvature tensor and the four-dimensional para-hyper-Kähler symmetric space (N^4, g_{PHK}) has the same signature and curvature tensor, the four-dimensional factor is isometric to (N^4, g_{PHK}). Finally, the metric g is Ricci-flat since g_{PHK} is Ricci-flat. $\qquad\square$

Example 4.3.8 The following examples define half-flat normalised SU(1, 2)-structures with $\omega^{t_1} = 0$ in a standard basis. None of the examples is flat. Thus, the four structures are equivalent as SO(2, 4)-structures due to Proposition 4.3.7, but the examples show that the geometry of the reduction to SU(1, 2) is not as rigid.

$$\omega = \omega_2, \quad \rho = e^{12}f^3 + \sqrt{2}e^{13}f^2 + e^1 f^{23} + e^{23}f^1 - e^3 f^{12} + \sqrt{2}f^{123},$$
$$g = -(e^2)^2 - (f^2)^2 + 2e^1 \cdot e^3 - 2\sqrt{2}e^1 \cdot f^3 + 2\sqrt{2}e^3 \cdot f^1 - 2f^1 \cdot f^3,$$
$$\text{(Ricci-flat pseudo-Kähler since } d\omega = 0, d\hat{\rho} = 0);$$

$$\omega = \omega_3, \quad \rho = e^{123} + e^{12}f^3 + e^{13}f^2 + e^1 f^{12} - 2e^1 f^{23} + e^2 f^{13} - e^3 f^{12},$$
$$g = -(e^2)^2 - 2(f^2)^2 + 2e^1 \cdot f^1 + 2e^1 \cdot f^3 + 2e^2 \cdot f^2 - 2e^3 \cdot f^1 - 2f^1 \cdot f^3,$$
$$(d\omega \neq 0, J_\rho \text{ integrable since } d\hat{\rho} = 0);$$

$$\omega = \omega_4, \quad \rho = \beta e^{12}f^3 - \beta e^{13}f^2 + \beta e^1 f^{23} + \frac{\beta+1}{\beta^3}e^{23}f^1 + \frac{\beta^4 - \beta - 1}{\beta^3}e^2 f^{13}$$
$$- \beta e^3 f^{12} - (\beta^2 + 2\beta)f^{123}, \qquad (d\omega \neq 0, d\hat{\rho} \neq 0),$$
$$g = -\frac{1}{\beta^2}(e^2)^2 - \beta^2 (f^2)^2 + 2\beta^2 e^1 \cdot f^3$$
$$- \frac{2}{\beta^2(\beta+1)}e^3 \cdot f^1 - \frac{2(\beta^4 + \beta + 1)}{\beta^2}f^1 \cdot f^3;$$

$$\omega = \omega_5, \quad \rho = e^{12}f^3 + e^{13}f^2 - e^1 f^{23} + e^{23}f^1 - e^3 f^{12} + f^{123}, \qquad (d\omega \neq 0, d\hat{\rho} \neq 0),$$
$$g = -(e^2)^2 - 2(f^2)^2 + 2e^1 \cdot e^3 + 2e^2 \cdot f^2 + 2f^1 \cdot f^3.$$

Example 4.3.9 Moreover, we give examples of half-flat normalised SL(3, $\mathbb{R}$)-structures with $\omega^{t_1} = 0$. Again, none of the structures is flat.

$$\omega = \omega_2, \quad \rho = \sqrt{2}(e^1 f^{23} + e^{23}f^1), \qquad (d\omega = 0, d\hat{\rho} = 0),$$
$$g = 2e^1 \cdot e^3 - 2e^2 \cdot f^2 - 2f^1 \cdot f^3;$$

$$\omega = \omega_3, \quad \rho = \sqrt{2}(e^{12}f^3 + e^{13}f^2 + e^1 f^{12} - e^3 f^{12}), \qquad (d\omega \neq 0, d\hat{\rho} \neq 0),$$

$$g = -2\,(e^1)^2 + 2\,e^1 \cdot e^3 - 2\,e^1 \cdot f^3 + 2\,e^2 \cdot f^2 - 2f^1 \cdot f^3;$$

$$\omega = \omega_4, \quad \rho = -\sqrt{2\beta + 2}\,(e^{12}f^3 - e^1 f^{23} + e^2 f^{13} - e^3 f^{12}), \qquad (d\omega \neq 0,\, d\hat{\rho} \neq 0),$$

$$g = -2\,(f^2)^2 + 2\,e^1 \cdot e^3 + 2\,e^1 \cdot f^3 + 2\,e^2 \cdot f^2 - 2\,e^3 \cdot f^1 - (2\beta + 4)f^1 \cdot f^3;$$

$$\omega = \omega_5, \quad \rho = \sqrt{2}\,(e^{123} + f^{123}), \qquad (d\omega \neq 0,\, d\hat{\rho} = 0),$$

$$g = 2\,e^1 \cdot f^3 + 2\,e^2 \cdot f^2 + 2\,e^3 \cdot f^1.$$

4.3.3 Solving the Evolution Equations on $H_3 \times H_3$

Due to the preparatory work of the Lemmas 4.3.1 and 4.3.4, it turns out to be possible to explicitly evolve every half-flat structure on $\mathfrak{h}_3 \oplus \mathfrak{h}_3$ without integrating.

Proposition 4.3.10 *Let (ω_0, ρ_0) be any half-flat $H^{\varepsilon,\tau}$-structure on $\mathfrak{h}_3 \oplus \mathfrak{h}_3$ with $\omega_0^{\mathfrak{k}_1} = 0$. Then, the solution of the evolution equations (4.29) is affine linear in the sense that*

$$\sigma(t) = \sigma_0 + t\,d\hat{\rho}_0, \qquad \rho(t) = \rho_0 + t\,d\omega_0 \tag{4.38}$$

and is well-defined for all $t \in \mathbb{R}$.

Proof Let $\{e_1, \ldots, f_3\}$ be a standard basis such that ω_0 is in one of the normal forms $\omega_2, \ldots, \omega_5$ of Lemma 4.3.4 which satisfy $\omega_0^{\mathfrak{k}_1} = 0$. By Lemma 4.3.1 and the second evolution equation, we know that there is a function $y(t)$ with $y(0) = 0$ such that

$$\sigma(t) = \sigma_0 + y(t)e^{12}f^{12} = \frac{1}{2}\omega_0^2 + y(t)e^{12}f^{12}.$$

For each of the four normal forms, the unique two-form $\omega(t)$ with $\frac{1}{2}\omega(t)^2 = \sigma(t)$ and $\omega(0) = \omega_0$ is

$$\omega(t) = \omega_0 - y(t)e^1 f^1.$$

However, the two-form $e^1 f^1$ is closed such that the exterior derivative $d\omega(t) = d\omega_0$ is constant. Therefore, we have $\rho(t) = \rho_0 + t\,d\omega_0$ by the first evolution equation. Moreover, the two-form $\omega(t)$ is stable for all $t \in \mathbb{R}$ since it holds $\phi(\omega(t)) = \phi(\omega_0)$ for each of the normal forms and for all $t \in \mathbb{R}$. It remains to show that $d\hat{\rho}(t)$ is constant in all four cases which implies that the function $y(t)$ is linear by the second evolution equation.

As explained in Sect. 4.3.2, it is easy to write down, for each normal form ω_0 separately, all compatible, closed three-forms ρ_0, which depend on nine parameters. For $\rho(t) = \rho_0 + t\,d\omega_0$, we verify with the help of a computer that $\lambda(\rho(t)) = \lambda(\rho_0)$ is constant such that $\rho(t)$ is stable for all $t \in \mathbb{R}$ since ρ_0 is stable. When we also calculate $J_{\rho(t)}$ and $\hat{\rho}(t) = J^*_{\rho(t)}\rho(t)$, it turns out in all four cases that $d\hat{\rho}(t)$ is constant. This finishes the proof.   □

We cannot expect that this affine linear evolution of spaces which have one-dimensional holonomy, due to Proposition 4.3.7, yields metrics with full holonomy G_2^*. Indeed, due to the following result the geometry does not change significantly compared to the six-manifold.

Corollary 4.3.11 *Let (ω_0, ρ_0) be a half-flat $H^{\varepsilon,\tau}$-structure on $\mathfrak{h}_3 \oplus \mathfrak{h}_3$ with $\omega_0^{\mathfrak{e}_1} = 0$ and let g_φ be the Ricci-flat metric induced by the parallel stable three-form φ on $M \times \mathbb{R}$ defined by the solution (4.38) of the evolution equations with initial value (ω_0, ρ_0). Then, the pseudo-Riemannian manifold $(M \times \mathbb{R}, g_\varphi)$ is either flat or isometric to the product of the four-dimensional para-hyper-Kähler symmetric space (N^4, g_{PHK}) and a three-dimensional flat factor.*

Proof By formula (2.24), the metric g_φ is determined by the time-dependent metric $g(t)$ induced by $(\omega(t), \rho(t))$. All assertions follow from the analysis of the curvature of g_φ completely analogous to the proof of Proposition 4.3.7.   □

The situation changes completely when we consider the first normal form ω_1 of Lemma 4.3.4.

Proposition 4.3.12 *Let (ω_0, ρ_0) be any normalised half-flat $H^{\varepsilon,\tau}$-structure on $\mathfrak{h}_3 \oplus \mathfrak{h}_3$ with $\omega_0^{\mathfrak{e}_1} \neq 0$. There is always a standard basis $\{e_1, \ldots, f_3\}$ such that $\omega_0 = e^1 f^1 + e^2 f^2 + e^3 f^3$. In such a basis, we define $(\omega(x), \rho(x))$ by*

$$\rho(x) = \rho_0 + x(e^{12}f^3 - e^3 f^{12}),$$

$$\omega(x) = 2\,(\varepsilon\kappa(x))^{-\frac{1}{2}}\left(\frac{1}{4}\varepsilon\kappa(x)\,e^1 f^1 + \frac{1}{4}\varepsilon\kappa(x)\,e^2 f^2 + e^3 f^3\right),$$

where $\kappa(x)\,(e^{123}f^{123})^{\otimes 2} = \lambda(\rho(x))$. Furthermore, let I be the maximal interval containing zero such that the polynomial $\kappa(x)$ of order four does not vanish for any $x \in I$. The parallel stable three-form (4.9) on $M \times I$ obtained by evolving (ω_0, ρ_0) along the Hitchin flow (4.29) is

$$\varphi = \frac{1}{2}\sqrt{\varepsilon\kappa(x)}\,\omega(x) \wedge dx + \rho(x).$$

The metric induced by φ, which has holonomy contained in $G^{\varepsilon,\tau}$, is by (4.10) given as

$$g_\varphi = g(x) - \frac{1}{4}\kappa(x)dx^2, \tag{4.39}$$

where $g(x)$ denotes the metric associated to $(\omega(x), \rho(x))$ via (2.6) and (2.17). The variable x is related to the parameter t of the Hitchin flow by the ordinary differential equation (4.42).

Proof Since $\omega_0^{t_1} \neq 0$, we can always choose a standard basis such that $\omega_0 = e^1 f^1 + e^2 f^2 + e^3 f^3$ is in the first normal form of Lemma 4.3.4. Then ρ_0 is of the form (4.32).

Moreover, by Lemma 4.3.1, there is a function $y(t)$ which is defined on an interval containing zero and satisfies $y(0) = 0$ such that the solution of the second evolution equation can be written

$$\sigma(t) = \sigma_0 + y(t)e^{12}f^{12}.$$

The unique $\omega(t)$ that satisfies $\omega(0) = \omega_0$ and $\frac{1}{2}\omega(t)^2 = \sigma(t)$ for all t is

$$\omega(t) = \sqrt{1 - y(t)}\, e^1 f^1 + \sqrt{1 - y(t)}\, e^2 f^2 + \frac{1}{\sqrt{1 - y(t)}}\, e^3 f^3.$$

Since

$$d\omega(t) = \frac{1}{\sqrt{1 - y(t)}}(e^{12}f^3 - e^3 f^{12}), \tag{4.40}$$

there is another function $x(t)$ with $x(0) = 0$ such that the solution of the first evolution equation can be written

$$\rho(t) = \rho_0 + x(t)(e^{12}f^3 - e^3 f^{12}). \tag{4.41}$$

This three-form is compatible with $\omega(t)$ for all t, as one can easily see from (4.32). Furthermore, the solution is normalised by Theorem 4.1.3, which implies

$$\sqrt{\varepsilon\lambda(\rho(t))} = \phi(\rho(t)) = 2\phi(\omega(t)) = -2\sqrt{1 - y(t)}\, e^{123}f^{123}.$$

Hence, we can eliminate $y(t)$ by

$$y(t) = 1 - \frac{1}{4}\varepsilon\kappa(x(t)).$$

We remark that the normalisation of $\rho_0 = \rho(0)$ corresponds to $\kappa(0) = 4\varepsilon$. Comparing (4.40) and (4.41), the evolution equations are equivalent to the single ordinary differential equation

$$\dot{x} = \frac{2}{\sqrt{\varepsilon\kappa(x(t))}} \tag{4.42}$$

for the only remaining parameter $x(t)$. In fact, we do not need to solve this equation in order to compute the parallel $G_2^{(*)}$-form when we substitute the coordinate t by x via the local diffeomorphism $x(t)$ satisfying $dt = \frac{1}{2}\sqrt{\varepsilon\kappa(x(t))}\, dx$. Inserting all substitutions into the formulas (4.9) and (4.10) for the stable three-form φ on $M \times I$

and the induced metric g_φ, all assertions of the proposition follow immediately from Theorem 4.1.3.   □

Example 4.3.13 The invariant $\kappa(x)$ and the induced metric $g(x)$ for the three explicit half-flat structures of Example 4.3.5 are the following.

If (ω_0, ρ_0) is the SU(3)-structure (4.33), it holds

$$\kappa(x) = (x - \sqrt{2})^3(x + \sqrt{2}), \qquad\qquad I = (-\sqrt{2}, \sqrt{2}),$$

$$g(x) = (1 - \frac{1}{2}\sqrt{2}x)\left((e^1)^2 + (e^2)^2 - 4\kappa(x)^{-1}(e^3)^2 + (e^4)^2 + (e^5)^2 - 4\kappa(x)^{-1}(e^6)^2\right)$$

$$+ \sqrt{2}x(1 - \frac{1}{2}\sqrt{2}x)\left(e^1 \cdot e^4 + e^2 \cdot e^5 + 4\kappa(x)^{-1}e^3 \cdot e^6\right).$$

If (ω_0, ρ_0) is the SU(1,2)-structure (4.34), we have

$$\kappa(x) = (x - \sqrt{2})(x + \sqrt{2})^3, \qquad\qquad I = (-\sqrt{2}, \sqrt{2}),$$

$$g(x) = (1 + \frac{1}{2}\sqrt{2}x)\left((e^1)^2 - (e^2)^2 + 4\kappa(x)^{-1}(e^3)^2 + (e^4)^2 - (e^5)^2 + 4\kappa(x)^{-1}(e^6)^2\right)$$

$$- \sqrt{2}x(1 + \frac{1}{2}\sqrt{2}x)\left(e^1 \cdot e^4 + e^2 \cdot e^5 + 4\kappa(x)^{-1}e^3 \cdot e^6\right).$$

And for the SL(3, $\mathbb{R}$)-structure (4.35), it holds

$$\kappa(x) = (2 + x^2)^2, \qquad\qquad I = \mathbb{R},$$

$$g(x) = (2 + x^2)\left(e^1 \cdot e^4 + e^2 \cdot e^5\right)$$

$$+ 4(2 - x^2)\kappa(x)^{-1}e^3 \cdot e^6 + 4\sqrt{2}\,x\kappa(x)^{-1}\left((e^3)^2 - (e^6)^2\right).$$

Theorem 4.3.14 *Let $(\omega(x), \rho(x))$ be the solution of the Hitchin flow with one of the three half-flat structures (ω_0, ρ_0) of Example 4.3.5 as initial value (see Proposition 4.3.12 for the explicit solution and Example 4.3.13 for the corresponding metric $g(x)$, defined for $x \in I$).*

Then, the holonomy of the metric g_φ on $M \times I$ defined by formula (4.39) equals G_2 for the SU(3)-structure (ω_0, ρ_0) and G_2^* for the other two structures.

Moreover, restricting the eight-parameter family of half-flat structures given by (4.32) to a small neighbourhood of the initial value (ρ_0, ω_0) yields in each case an eight-parameter family of metrics of holonomy equal to G_2 or G_2^*.

Proof For all three cases, we can apply the transformation (4.36) and calculate the curvature R of the metric g_φ defined by (4.39). Carrying this out with the package "tensor" contained in Maple 10, we obtained that the rank of the curvature viewed as endomorphism on two-vectors is 14. This implies that the holonomy of g_φ in fact equals G_2 or G_2^*.

The assertion for the eight-parameter family is an immediate consequence. Indeed, by construction, the rank of the curvature endomorphism is bounded from above by 14 and being of maximal rank is an open condition. □

To conclude this section we address the issue of completeness and use the Riemannian family in Example 4.3.13 and Corollary 4.1.9 to construct a complete conformally parallel G_2-metric on $\mathbb{R} \times (\Gamma\backslash H_3 \times H_3)$.

Example 4.3.15 Let H_3 be the Heisenberg group and $N = \Gamma\backslash H_3 \times H_3$ be a compact nilmanifold given by a lattice Γ. Let us denote by $x : I \to (-\sqrt{2}, \sqrt{2})$ the maximal solution to the equation

$$\dot{x}(t) = \frac{2}{\sqrt{(\sqrt{2} - x(t))^3 (x(t) + \sqrt{2})}},$$

with initial condition $x(0) = 0$, defining the t-dependent family of Riemannian metrics

$$g_t = \frac{\sqrt{2} - x(t)}{\sqrt{2}} \left((e^1)^2 + (e^2)^2 + (e^4)^2 + (e^5)^2 \right) + x(t) \left(\sqrt{2} - x(t) \right) \left(e^1 \cdot e^4 + e^2 \cdot e^5 \right)$$

$$+ \frac{2\sqrt{2}}{(\sqrt{2} - x(t))^2 (x(t) + \sqrt{2})} \left((e^3)^2 + (e^6)^2 \right) - \frac{4x(t)}{(\sqrt{2} - x(t))^2 (x(t) + \sqrt{2})} e^3 \cdot e^6.$$

If $\varphi : \mathbb{R} \to I$ is a diffeomorphism, then the metric

$$dr^2 + \frac{1}{\varphi'(r)^2} g_{\varphi(r)}$$

is globally conformally parallel G_2 and geodesically complete.

4.4 Special Geometry of Real Forms of the Symplectic SL(6, ℂ)-Module $\wedge^3\mathbb{C}^6$

Homogeneous projective special Kähler manifolds of semisimple groups with possibly indefinite metric and compact stabiliser were classified in [4]. This includes the case of manifolds with (positive or negative) definite metrics, for which the stabiliser is automatically compact. Projective special Kähler manifolds with negative definite metric play an important role in supergravity and string theory. The space of local deformations of the complex structure of a Calabi Yau three-fold, for instance, is an example of a projective special Kähler manifold with negative definite metric. As a particular result of the classification [4], there is an interesting one-to-one correspondence between complex simple Lie algebras 𝖑 of type A, B, D, E, F and G and homogeneous projective special Kähler manifolds of semisimple groups with negative definite metric. The resulting spaces are certain Hermitian symmetric spaces of non-compact type. The homogeneous projective special Kähler manifold associated to the complex simple Lie algebra of type E_6, for instance, is precisely the Hermitian symmetric space $SU(3, 3)/ S(U(3) \times U(3))$.

Under the above assumptions, the homogeneous projective special Kähler manifold G/K is realised as an open orbit of a real semisimple group G acting on a smooth projective algebraic variety $X \subset P(V)$, where V is the complexification of a real symplectic module V_0 of G and the cone $C(X) := \{v \in V \mid \pi(v) \in X\} \subset V$ over X is Lagrangian. Here $\pi : V \setminus \{0\} \to P(V)$ denotes the canonical projection. In fact, $C(X)$ is the orbit of the highest weight vector of the $G^{\mathbb{C}}$-module V under the complexified group $G^{\mathbb{C}}$. In the case $G/K = SU(3, 3)/ S(U(3) \times U(3))$ the symplectic module is given by $V = \wedge^3 \mathbb{C}^6$. It was shown in [14] that the real symplectic G-module V_0 always admits a homogeneous quartic invariant λ, which is related to the hyper-Kähler part of the curvature tensor of a symmetric quaternionic Kähler manifold associated to the given complex simple Lie algebra $\mathfrak{l}$. Moreover, the level sets $\{\lambda = c\}$ are proper affine hyperspheres for $c \neq 0$ and the affine special Kähler manifold M underlying the homogeneous projective special Kähler manifold $\bar{M} = G/K$ can be realised as one of the open orbits of $\mathbb{R}^* \cdot G$ on V_0 [14].

In the following we shall describe all real forms (G, V_0) of the $SL(6, \mathbb{C})$-module $V = \wedge^3 \mathbb{C}^6$ and study the affine special geometry of the corresponding open orbits of $\mathbb{R}^* \cdot G$. As a consequence, we obtain a list of projective special Kähler manifolds, which admit a transitive action of a real form of $SL(6, \mathbb{C})$ by automorphisms of the special Kähler structure. Besides the unique stationary compact example

$$SU(3, 3)/ S(U(3) \times U(3)),$$

we obtain the homogeneous projective special Kähler manifolds

$$SU(3, 3)/ S(U(2, 1) \times U(1, 2)), \quad SU(5, 1)/ S(U(3) \times U(2, 1))$$
$$\text{and} \quad SL(6, \mathbb{R})/ (U(1) \cdot SL(3, \mathbb{C})),$$

which are symmetric spaces with indefinite metrics and non-compact stabiliser. The Hermitian signature of the metric is $(4, 5)$, $(6, 3)$ and $(3, 6)$, respectively. The latter result $(3, 6)$ corrects Proposition 7 in [76], according to which the Hermitian signature of the underlying affine special Kähler manifold $GL^+(6, \mathbb{R})/ SL(3, \mathbb{C})$ is $(1, 9)$. The correct Hermitian signature of the affine special Kähler manifold is $(4, 6)$.

Finally, we find that one of the two open orbits of $SL(6, \mathbb{R})$ on $\wedge^3 \mathbb{R}^6$ carries affine special para-Kähler geometry, the geometry of $N = 2$ vector multiplets on Euclidian rather than Minkowskian space-time [44]. The corresponding homogeneous projective special para-Kähler manifold is the symmetric space

$$SL(6, \mathbb{R})/ S(GL(3, \mathbb{R}) \times GL(3, \mathbb{R})).$$

4.4.1 The Symplectic SL(6, $\mathbb{C}$)-Module $V = \wedge^3\mathbb{C}^6$ and Its Lagrangian Cone $C(X)$ of Highest Weight Vectors

We consider the 20-dimensional irreducible SL(6, $\mathbb{C}$)-module $V = \wedge^3\mathbb{C}^6$ equipped with a generator ν of the line $\wedge^6\mathbb{C}^6$. The choice of ν determines an SL(6, $\mathbb{C}$)-invariant symplectic form Ω, which given by

$$\Omega(v, w)\nu = v \wedge w, \quad v, w \in V. \tag{4.43}$$

The highest weight vectors in V are precisely the non-zero decomposable three-vectors. They form a cone $C(X) \subset V$ over a smooth projective variety $X \subset P(V)$, namely the Grassmannian $Gr_3(\mathbb{C}^6)$ of complex three-planes in $\mathbb{C}^6$. The group SL(6, $\mathbb{C}$) acts transitively on the cone $C(X)$ and, hence, on the compact variety

$$X \cong SL(6, \mathbb{C})/P \cong SU(6)/S(U(3) \times U(3)),$$

where $P = SL(6, \mathbb{C})_x \subset SL(6, \mathbb{C})$ is the stabiliser of a point $x \in X$ (a parabolic subgroup).

Proposition 4.4.1 *The cone $C(X) = \{v \in \wedge^3\mathbb{C}^6 \setminus \{0\} \mid v$ is decomposable$\} \subset V$ is Lagrangian.*

Proof Let $(e_1, \ldots, e_6)$ be a basis of $\mathbb{C}^6$ and put $p = e_{123}$. Then

$$T_pC(X) = \text{span}\{e_{ijk} \mid \#\{i, j, k\} \cap \{1, 2, 3\} \geq 2\}$$

is ten-dimensional and is clearly totally isotropic with respect to Ω. $\qquad\square$

4.4.2 Real Forms (G, V_0) of the Complex Module $(SL(6, \mathbb{C}), V)$

Let G be a real form of the complex Lie group SL(6, $\mathbb{C}$). There exists a G-invariant real structure τ on $V = \wedge^3\mathbb{C}^6$ if and only if $G = SL(6, \mathbb{R})$, SU(3, 3), SU(5, 1). In the first case τ is simply complex conjugation with respect to $V_0 = \wedge^3\mathbb{R}^6$. In order to describe the real structure in the other two cases, we first endow $\mathbb{C}^6$ with the standard SU(p, q)-invariant pseudo-Hermitian form $\langle \cdot, \cdot \rangle$. The pseudo-Hermitian form $\langle \cdot, \cdot \rangle$ on $\mathbb{C}^6$ induces an SU(p, q)-invariant pseudo-Hermitian form γ on V such that

$$\gamma(v_1 \wedge v_2 \wedge v_3, w_1 \wedge w_2 \wedge w_3) = \det(\langle v_i, w_j \rangle), \tag{4.44}$$

for all $v_1, \ldots, w_3 \in \mathbb{C}^6$. Then we define an $SU(p,q)$-invariant anti-linear map $\tau : V \to V$ by the equation

$$\tau := \sqrt{-1}\gamma^{-1} \circ \Omega.$$

Notice that $\Omega : V \to V^*, v \mapsto \Omega(\cdot, v)$ is linear, whereas $\gamma : V \to V^*, v \mapsto \gamma(\cdot, v)$ and $\gamma^{-1} : V^* \to V$ are anti-linear.

Proposition 4.4.2 *The anti-linear map τ is an $SU(p,q)$-invariant real structure on $V = \wedge^3 \mathbb{C}^6$ if and only if $p - q \equiv 0$ (mod 4). In that case, the $SU(p,q)$-invariant pseudo-Hermitian form $\gamma = \sqrt{-1}\Omega(\cdot, \tau \cdot)$ on V has signature $(10, 10)$. Otherwise, τ is an $SU(p,q)$-invariant quaternionic structure on V.*

Proof We present the calculations in the relevant cases $(p, q) = (3, 3)$ and $(p, q) = (5, 1)$. The calculations in the other cases are similar.

Case $(p, q) = (3, 3)$. Let $(e_1, e_2, e_3, f_1, f_2, f_3)$ be a unitary basis of $(\mathbb{C}^6, \langle \cdot, \cdot \rangle) = \mathbb{C}^{3,3}$, such that $\langle e_i, e_i \rangle = -\langle f_i, f_i \rangle = 1$. We consider the following basis of V:

$$(e_{123}, e_1 \wedge f_{12}, e_1 \wedge f_{13}, e_1 \wedge f_{23}, e_2 \wedge f_{12}, e_2 \wedge f_{13}, e_2 \wedge f_{23}, e_3 \wedge f_{12}, e_3 \wedge f_{13}, e_3 \wedge f_{23},$$

$$f_{123}, e_{23} \wedge f_3, -e_{23} \wedge f_2, e_{23} \wedge f_1, -e_{13} \wedge f_3, e_{13} \wedge f_2, -e_{13} \wedge f_1, e_{12} \wedge f_3, -e_{12} \wedge f_2, e_{12} \wedge f_1).$$

With respect to that basis and $v = e_{123} \wedge f_{123}$ we have

$$\gamma = \begin{pmatrix} \mathbb{1}_{10} & 0 \\ 0 & -\mathbb{1}_{10} \end{pmatrix}, \quad \Omega = \begin{pmatrix} 0 & \mathbb{1}_{10} \\ -\mathbb{1}_{10} & 0 \end{pmatrix}, \quad \tau = \sqrt{-1}\gamma^{-1} \circ \Omega = \sqrt{-1}\begin{pmatrix} 0 & \mathbb{1}_{10} \\ \mathbb{1}_{10} & 0 \end{pmatrix}.$$
$$(4.45)$$

This implies $\tau^2 = \mathrm{Id}$, since τ is anti-linear.

Case $(p, q) = (5, 1)$. Let $(e_1, \ldots, e_5, f)$ be a unitary basis of $(\mathbb{C}^6, \langle \cdot, \cdot \rangle) = \mathbb{C}^{5,1}$, such that $\langle e_i, e_i \rangle = -\langle f, f \rangle = 1$. With respect to the basis

$$(e_{123}, e_{124}, e_{125}, e_{134}, e_{135}, e_{145}, e_{234}, e_{235}, e_{245}, e_{345}, \qquad\qquad (4.46)$$

$$e_{45} \wedge f, -e_{35} \wedge f, e_{34} \wedge f, e_{25} \wedge f, -e_{24} \wedge f, e_{23} \wedge f, -e_{15} \wedge f, e_{14} \wedge f, -e_{13} \wedge f, e_{12} \wedge f)$$

of V and $v = e_{12345} \wedge f$ we have again the formulas (4.45) and $\tau^2 = \mathrm{Id}$. $\qquad\square$

4.4.3 Classification of Open G-Orbits on the Grassmannian X and Corresponding Special Kähler Manifolds

For each of the real forms (G, V_0) obtained in the previous section, we will now describe all open orbits of the real simple Lie group G on the Grassmannian $X = Gr_3(\mathbb{C}^6) = \{E \subset \mathbb{C}^6 \text{ a three-dimensional subspace}\} \hookrightarrow P(V), E \mapsto \wedge^3 E$. We

Table 4.1 Homogeneous projective special Kähler manifolds $\bar{M} = G/H$ of real simple groups G of type A_5. Notice that $\dim_{\mathbb{C}} \bar{M} = 9$

G/H	Hermitian signature
SU(3, 3)/ S(U(3) × U(3))	(0,9)
SU(3, 3)/ S(U(2, 1) × U(1, 2))	(4,5)
SU(5, 1)/ S(U(3) × U(2, 1))	(6,3)
SL(6, $\mathbb{R}$)/ (U(1) · SL(3, $\mathbb{C}$))	(3,6)

will also describe the projective special Kähler structure of these orbits $\bar{M} \subset P(V)$ and the (affine) special Kähler structure of the corresponding cones $M = C(\bar{M}) \subset V$. The resulting homogeneous projective special Kähler manifolds are listed in Table 4.1. Let us first recall some definitions and constructions from special Kähler geometry.

Basic Facts About Special Kähler Manifolds

Definition 4.4.3 A *special Kähler manifold* (M, J, g, ∇) is a (pseudo-)Kähler manifold (M, J, g) endowed with a flat torsion-free connection ∇ such that $\nabla \omega = 0$ and $d^\nabla J = 0$, where $d^\nabla J$ is the exterior covariant derivative of the vector valued one-form J.

A *conical special Kähler manifold* (M, J, g, ∇, ξ) is a special Kähler manifold (M, J, g, ∇) endowed with a timelike or a spacelike vector field ξ such that $\nabla \xi = D\xi = \mathrm{Id}$, where D is the Levi-Civita connection. The vector field ξ is called *Euler vector field*.

The vector fields ξ and $J\xi$ generate a free holomorphic action of a two-dimensional Abelian Lie algebra. If the action can be integrated to a free holomorphic $\mathbb{C}^*$-action such that the quotient map $M \to \bar{M} := M/\mathbb{C}^*$ is a holomorphic submersion, then $\bar{M}$ is called a *projective special Kähler manifold*. We will see now that $\bar{M}$ carries a canonical (pseudo-)Kähler metric $\bar{g}$ compatible with the induced complex structure J on $\bar{M} = M/\mathbb{C}^*$. Multiplying the metric g with -1 if necessary, we can assume that $g(\xi, \xi) > 0$. Then $S = \{p \in M \mid g(\xi(p), \xi(p)) = 1\}$ is a smooth hypersurface invariant under the isometric S^1-action generated by the Killing vector field $J\xi$ and we can recover $\bar{M}$ as the base of the circle bundle $S \to \bar{M} = S/S^1$. Then $\bar{M}$ carries a unique pseudo-Riemannian metric $\bar{g}$ such that $S \to \bar{M}$ is a Riemannian submersion. $(\bar{M}, J, \bar{g})$ is in fact the Kähler quotient of (M, J, g) by the S^1-action generated by the Hamiltonian Killing vector field $J\xi$.

Next we explain the extrinsic construction of special Kähler manifolds from [6]. Let (V, Ω) be a complex symplectic vector space of dimension $2n$ endowed with a real structure τ such that

$$\overline{\Omega(v, w)} = \Omega(\tau v, \tau w), \quad \text{for all} \quad v, w \in V. \tag{4.47}$$

Then the pseudo-Hermitian form

$$\gamma := \sqrt{-1}\Omega(\cdot, \tau\cdot) \tag{4.48}$$

has signature (n, n).

Definition 4.4.4 A holomorphic immersion $\phi : M \to V$ from an n-dimensional complex manifold (M, J) into V is called

(i) *nondegenerate* if $\phi^*\gamma$ is nondegenerate,
(ii) *Lagrangian* if $\phi^*\Omega = 0$ and
(iii) *conical* if $\phi(p) \in d\phi(T_pM)$ and $\gamma(\phi(p), \phi(p)) \neq 0$ for all $p \in M$.

Theorem 4.4.5 ([6])

(i) *Any nondegenerate Lagrangian immersion $\phi : M \to V$ induces on the complex manifold (M, J) the structure of a special Kähler manifold (M, J, g, ∇), where $g = \operatorname{Re} \phi^*\gamma$ and ∇ is determined by the condition $\nabla\phi^*\alpha = 0$ for all $\alpha \in V^*$ which are real valued on V^τ.*
(ii) *Any conical nondegenerate Lagrangian immersion $\phi : M \to V$ induces on (M, J) the structure of a conical special Kähler manifold (M, J, g, ∇, ξ). The vector field ξ is determined by the condition $d\phi \, \xi(p) = \phi(p)$.*

The Case $G = \mathrm{SL}(6, \mathbb{R})$

Using the complex conjugation $\tau : v \mapsto \bar{v}$ on $\mathbb{C}^6$ we can decompose X into G-invariant real algebraic subvarieties $X_{(k)} = X_{(k)}(\tau) := \{E \in X \mid \dim(E \cap \bar{E}) = k\} \subset X$, where $E \subset \mathbb{C}^6$ runs through all three-dimensional subspaces and $k \in \{0, 1, 2, 3\}$. Notice that only $X_{(0)} \subset X$ is open.

Proposition 4.4.6 *The group $\mathrm{SL}(6, \mathbb{R})$ acts transitively on the open real subvariety $X_{(0)} = \{E \in X \mid E \cap \bar{E} = 0\} \subset X$.*

Proof Given bases (e_1, e_2, e_3), (e'_1, e'_2, e'_3) of $E, E' \in X_{(0)}$, respectively, let φ be the linear transformation, which maps the basis $(e_1, e_2, e_3, \bar{e}_1, \bar{e}_2, \bar{e}_3)$ of $\mathbb{C}^6$ to the basis $(e'_1, e'_2, e'_3, \bar{e}'_1, \bar{e}'_2, \bar{e}'_3)$ of $\mathbb{C}^6$. Then $\varphi \in \mathrm{SL}(6, \mathbb{R})$ and $\varphi E = E'$. $\square$

Theorem 4.4.7 *The group $\mathrm{SL}(6, \mathbb{R})$ has a unique open orbit $X_{(0)} \cong \mathrm{SL}(6, \mathbb{R})/ (\mathrm{U}(1) \cdot \mathrm{SL}(3, \mathbb{C}))$ on the highest weight variety $P(V) \supset X \cong \mathrm{Gr}_3(\mathbb{C}^6)$ of $V = \wedge^3\mathbb{C}^6$. The cone $V \supset M = C(X_{(0)}) \cong \mathrm{GL}^+(6, \mathbb{R})/ \mathrm{SL}(3, \mathbb{C})$ carries an $\mathrm{SL}(6, \mathbb{R})$-invariant special Kähler structure of Hermitian signature $(4, 6)$, which induces on $\bar{M} = X_{(0)}$ the structure of a homogeneous projective special Kähler manifold $\bar{M}$ of Hermitian signature $(3, 6)$.*

Proof Let $e_1, e_2, e_3 \in \mathbb{C}^6$ be three vectors which span a three-dimensional subspace $E \subset \mathbb{C}^6$ such that $E \in X_{(0)}$. Then $\mathbb{C}^6 = E \oplus \bar{E}$ and the tangent space of $M = C(X_{(0)})$ at $p = e_{123}$ is given by $T_pM = \wedge^3E \oplus \wedge^2E \wedge \bar{E}$. We choose the real generator

$v = \sqrt{-1}e_{123} \wedge \bar{e}_{123} \in \wedge^3\mathbb{R}^6$ and compute $\gamma = \sqrt{-1}\Omega(\cdot, \tau\cdot)$ on T_pM using the formula (4.43). The matrix of $\gamma|_{T_pM}$ with respect to the basis ($e_{123}, e_{12} \wedge \bar{e}_3, e_{13} \wedge \bar{e}_2, e_{23} \wedge \bar{e}_1, e_{12} \wedge \bar{e}_1, e_{13} \wedge \bar{e}_1, e_{12} \wedge \bar{e}_2, -e_{23} \wedge \bar{e}_3, e_{23} \wedge \bar{e}_2, e_{13} \wedge \bar{e}_3$) is given by

$$
\begin{pmatrix}
1 & 0 & 0 & 0 \\
0 & -\mathbb{1}_3 & 0 & 0 \\
0 & 0 & 0 & \mathbb{1}_3 \\
0 & 0 & \mathbb{1}_3 & 0
\end{pmatrix}.
\tag{4.49}
$$

This shows that γ has signature $(4, 6)$ on T_pM. Since $\mathrm{GL}^+(6, \mathbb{R})$ acts transitively on M and preserves the pseudo-Hermitian form γ up to a positive factor ($\mathrm{SL}(6, \mathbb{R})$ acts isometrically), the signature of the restriction of γ to M does not depend on the base point. This shows that the inclusion $M \subset V$ is a holomorphic conical nondegenerate Lagrangian immersion. By Theorem 4.4.5, it induces a conical special (pseudo-)Kähler structure (J, g, ∇, ξ) on M. It follows that the image $\bar{M} = \pi(M) = X_{(0)} \cong \mathrm{SL}(6, \mathbb{R})/(\mathrm{U}(1) \cdot \mathrm{SL}(3, \mathbb{C}))$ of M under the projection $\pi : V \setminus \{0\} \to P(V)$ is a homogeneous projective special Kähler manifold. The induced pseudo-Hermitian form $\bar{\gamma}$ on $T_{\pi(p)}\bar{M}$ has signature $(3, 6)$. The latter statement follows from formula $\bar{\gamma}(d\pi_pX, d\pi_pY) = \frac{\gamma(X,Y)}{\gamma(p,p)}$ for $X, Y \in T_pM \cap p^\perp \subset V$ (see [4]), since $\gamma(p, p) = 1$ for $p = e_{123}$. □

The Case $G = \mathrm{SU}(3, 3)$

Using the pseudo-Hermitian form $h = \langle \cdot, \cdot \rangle$ on $\mathbb{C}^6$ invariant under $G = \mathrm{SU}(3, 3)$ we can decompose the Grassmannian $X = Gr_3(\mathbb{C}^6)$ into the G-invariant real algebraic subvarieties $X_{(k)} = X_{(k)}(h) := \{E \in X \mid \mathrm{rk}(h|_E) = k\}$, $k \in \{0, 1, 2, 3\}$. Notice that only $X_{(3)} \subset X$ is open and that it can be decomposed further according to the possible signatures of $h|_E$:

$$X_{(s,t)} := \{E \in X \mid E \text{ has signature } (s, t)\},$$

where $(s, t) \in \{(3, 0), (2, 1), (1, 2), (0, 3)\}$.

Theorem 4.4.8 *The group* $\mathrm{SU}(3, 3)$ *has precisely four open orbits on the highest weight variety* $P(V) \supset X \cong Gr_3(\mathbb{C}^6)$ *of* $V = \wedge^3\mathbb{C}^6$, *namely* $X_{(3,0)}$, $X_{(2,1)}$, $X_{(1,2)}$ *and* $X_{(0,3)}$. *In all four cases the cone* $M = C(X_{(s,t)}) \subset V$ *carries an* $\mathrm{SU}(3, 3)$-*invariant special Kähler structure.*

$$C(X_{(3,0)}) \cong \mathbb{R}^* \cdot \mathrm{SU}(3, 3)/\mathrm{SU}(3) \times \mathrm{SU}(3)$$

has Hermitian signature $(1, 9)$. $C(X_{(0,3)}) \cong \mathbb{R}^* \cdot \mathrm{SU}(3, 3)/\mathrm{SU}(3) \times \mathrm{SU}(3)$ *has Hermitian signature* $(9, 1)$. *For* $\{s, t\} = \{2, 1\}$,

$$C(X_{(s,t)}) \cong \mathbb{R}^* \cdot \mathrm{SU}(3, 3)/\mathrm{SU}(2, 1) \times \mathrm{SU}(1, 2)$$

has Hermitian signature $(5, 5)$. In all cases, the conical special Kähler manifold $M = C(X_{(s,t)})$ induces on $\bar{M} = X_{(s,t)}$ the structure of a homogeneous projective special Kähler manifold $\bar{M}$. For $\{s, t\} = \{3, 0\}$,

$$\bar{M} = X_{(s,t)} \cong X_{(3,0)} \cong SU(3,3)/S(U(3) \times U(3))$$

has Hermitian signature $(0, 9)$, for $\{s, t\} = \{2, 1\}$,

$$\bar{M} = X_{(s,t)} \cong X_{(2,1)} \cong SU(3,3)/S(U(2,1) \times U(1,2))$$

has Hermitian signature $(4, 5)$. The special Kähler manifolds $C(X_{(s,t)})$ and $C(X_{(t,s)})$ are equivalent. In fact, they are related by a holomorphic ∇-affine anti-isometry, which induces a holomorphic isometry between the corresponding projective special Kähler manifolds.

Proof $X_{(3)} \subset X$ is Zariski open and is decomposed into the four open (in the standard topology) orbits $X_{(s,t)}$ of $G = SU(3, 3)$. Let $(e_1, e_2, e_3, f_1, f_2, f_3)$ be a unitary basis of $(\mathbb{C}^6, h)$, such that $\langle e_i, e_i \rangle = -\langle f_i, f_i \rangle = 1$. Then $X_{(3,0)}, X_{(2,1)}, X_{(1,2)}, X_{(0,3)}$ are the G-orbits of the lines generated by the elements $e_{123}, e_1 \wedge f_{12}, e_{12} \wedge f_1, f_{123} \in V$, respectively.

For $p = e_{123}$ and $M = C(X_{(3,0)})$, we calculate

$$T_p M = \text{span}\{e_{123}, f_i \wedge e_{jk}\}.$$

From (4.44) we see that the matrix of $\gamma|_{T_p M}$ with respect to the basis $(e_{123}, f_i \wedge e_{jk})$ is $\text{diag}(1, -\mathbb{1}_9)$. Therefore γ has signature $(1, 9)$ on $T_p M$. Since $\gamma(e_{123}, e_{123}) = 1$, the signature of the induced pseudo-Hermitian form $\bar{\gamma}$ on $T_p \bar{M}$ is $(0, 9)$.

For $p = e_{12} \wedge f_1$ and $M = C(X_{(2,1)})$, we obtain

$$T_p M = \text{span}\{e_1 \wedge f_{12}, e_2 \wedge f_{12}, e_2 \wedge f_{13}, e_1 \wedge f_{13}, e_{123}, e_{ij} \wedge f_1, e_{12} \wedge f_i\}.$$

The matrix of $\gamma|_{T_p M}$ with respect to the above basis is $\text{diag}(\mathbb{1}_5, -\mathbb{1}_5)$. Therefore γ has signature $(5, 5)$ on $T_p M$. We have $\gamma(e_{12} \wedge f_1, e_{12} \wedge f_1) = -1$ and the signature of the induced pseudo-Hermitian form $\bar{\gamma}$ on $T_p \bar{M}$ is $(4, 5)$.

The linear transformation which sends the vectors e_i to f_i and f_i to e_i induces a linear map $\varphi : V \to V$, which interchanges the cone $C(X_{(s,t)})$ with $C(X_{(t,s)})$ and maps γ to $-\gamma$. This shows that γ has signature $(9, 1)$ and $(5, 5)$ on $C(X_{(0,3)})$ and $C(X_{(1,2)})$, respectively. As a consequence, the induced pseudo-Hermitian form $\bar{\gamma}$ on $\bar{M} = X_{(0,3)}, X_{(1,2)}$, has still signature $(0, 9)$ and $(4, 5)$, respectively.

It follows from these calculations that the inclusion $M = C(X_{(s,t)}) \subset V$ is a holomorphic conical nondegenerate Lagrangian immersion. By Theorem 4.4.5, it induces a conical special (pseudo-)Kähler structure (J, g, ∇, ξ) on M and $\bar{M} = X_{(s,t)} \subset P(V)$ is a projective special Kähler manifold.

The above linear anti-isometry $\varphi : (V, \gamma) \to (V, \gamma)$ maps Ω to $-\Omega$, and, hence, preserves the real structure $\tau = \sqrt{-1}\gamma^{-1} \circ \Omega$. As a result, it maps the special Kähler structure (J, g, ∇) of $C(X_{(s,t)})$ to $(J', -g', \nabla')$, where (J', g', ∇') is the special Kähler structure of $C(X_{(t,s)})$. In particular, it induces a holomorphic isometry $X_{(s,t)} \cong X_{(t,s)}$. $\qquad\qquad\qquad\qquad\qquad\qquad\qquad\qquad\qquad\qquad\qquad\qquad\square$

The Case $G = \mathrm{SU}(5, 1)$

Let $h = \langle\cdot, \cdot\rangle$ be the standard pseudo-Hermitian form of signature $(5, 1)$ on $\mathbb{C}^6$, which is invariant under $G = \mathrm{SU}(5, 1)$. Let us fix a unitary basis $(e_1, \ldots, e_5, f)$ of $(\mathbb{C}^6, h)$, such that $\langle e_i, e_i\rangle = -\langle f, f\rangle = 1$. As in the previous subsection, $X = Gr_3(\mathbb{C}^6)$ is decomposed into the G-invariant real algebraic subvarieties $X_{(k)} = X_{(k)}(h)$, of which $X_{(3)} \subset X$ is Zariski open. $X_{(3)}$ is now the union of the two open G-orbits $X_{(3,0)}$ and $X_{(2,1)}$. $X_{(3,0)}$ is the orbit of the line $\mathbb{C}e_{123} \in P(V)$ and $X_{(2,1)}$ is the orbit of $\mathbb{C}e_{45} \wedge f \in P(V)$.

Theorem 4.4.9 *The group* $\mathrm{SU}(5, 1)$ *has precisely two open orbits on the highest weight variety* $P(V) \supset X \cong Gr_3(\mathbb{C}^6)$ *of* $V = \wedge^3\mathbb{C}^6$, *namely* $X_{(3,0)}$ *and* $X_{(2,1)}$. *In both cases the cone* $M = C(X_{(s,t)}) \subset V$ *carries an* $\mathrm{SU}(5, 1)$-*invariant special Kähler structure.*

$$C(X_{(3,0)}) \cong \mathbb{R}^* \cdot \mathrm{SU}(5, 1)/ \mathrm{SU}(3) \times \mathrm{SU}(2, 1)$$

has Hermitian signature $(7, 3)$.

$$C(X_{(2,1)}) \cong \mathbb{R}^* \cdot \mathrm{SU}(5, 1)/ \mathrm{SU}(3) \times \mathrm{SU}(2, 1)$$

has Hermitian signature $(3, 7)$. *In both cases, the conical special Kähler manifold* $M = C(X_{(s,t)})$ *induces on* $\bar{M} = X_{(s,t)}$ *the structure of a homogeneous projective special Kähler manifold* $\bar{M}$.

$$\bar{M} = X_{(3,0)} \cong \mathrm{SU}(5, 1)/ \mathrm{S}(\mathrm{U}(3) \times \mathrm{U}(2, 1))$$

and

$$\bar{M} = X_{(2,1)} \cong \mathrm{SU}(5, 1)/ \mathrm{S}(\mathrm{U}(3) \times \mathrm{U}(2, 1))$$

have both Hermitian signature $(6, 3)$. *The special Kähler manifolds* $C(X_{(3,0)})$ *and* $C(X_{(2,1)})$ *are equivalent. In fact, they are related by a holomorphic* ∇-*affine anti-isometry, which induces a holomorphic isometry between the corresponding projective special Kähler manifolds.*

Proof For $p = e_{123}$ and $M = C(X_{(3,0)})$, we have

$$T_pM = \mathrm{span}\{e_{123}, e_{234}, e_{235}, e_{134}, e_{135}, e_{124}, e_{125}, e_{12} \wedge f, e_{13} \wedge f, e_{23} \wedge f\}$$

and the restriction of γ to T_pM is represented by the matrix $\mathrm{diag}(\mathbb{1}_7, -\mathbb{1}_3)$ with respect to the above basis. This shows that the inclusion $M = C(X_{(3,0)}) \subset V$ is a holomorphic conical nondegenerate Lagrangian immersion. By Theorem 4.4.5, it induces a conical special (pseudo-)Kähler structure (J, g, ∇, ξ) of Hermitian signature $(7, 3)$ on M and $\bar{M} = X_{(3,0)} \subset P(V)$ is a projective special Kähler manifold of Hermitian signature $(6, 3)$. The anti-isometry relating $C(X_{(3,0)})$ and $C(X_{(2,1)})$ is induced by the linear map $\varphi : V \to V$ which has the matrix

$$\begin{pmatrix} 0 & \mathbb{1}_{10} \\ \mathbb{1}_{10} & 0 \end{pmatrix},$$

with respect to the basis (4.46).   □

4.4.4 The Homogeneous Projective Special Para-Kähler Manifold $\mathbf{SL(6, \mathbb{R}) / S(GL(3, \mathbb{R}) \times GL(3, \mathbb{R}))}$

Let us first briefly recall the necessary definitions and constructions from special para-Kähler geometry, see [44] for more details.

Basic Facts About Special Para-Kähler Manifolds

Definition 4.4.10 A *special para-Kähler manifold* (M, P, g, ∇) is a para-Kähler manifold (M, P, g) endowed with a flat torsion-free connection ∇ such that $\nabla \omega = 0$ and $d^\nabla P = 0$, where $\omega = g(\cdot, P\cdot)$.

A *conical special para-Kähler manifold* (M, P, g, ∇, ξ) is a special para-Kähler manifold
(M, P, g, ∇) endowed with a timelike or a spacelike vector field ξ such that $\nabla \xi = D\xi = \mathrm{Id}$, where D is the Levi-Civita connection.

It follows from the definition of a para-Kähler manifold that the eigenspaces of P are of the same dimension and involutive. An endomorphism field $P \in \Gamma(\mathrm{End}\, TM)$ with these properties is called a *para-complex structure* on M. The pair (M, P) is then called a *para-complex manifold*. A smooth map $f : (M, P_M) \to (N, P_N)$ between para-complex manifolds is called *para-holomorphic* if $df \circ P_M = P_N \circ df$. The skew-symmetry of P in the definition of a para-Kähler manifold implies that the eigenspaces of P are totally isotropic of dimension $n = \frac{1}{2} \dim M$. In particular, M is of even dimension $2n$ and g is of signature (n, n).

On any conical special para-Kähler manifold, the vector fields ξ and $P\xi$ generate a free para-holomorphic action of a two-dimensional Abelian Lie algebra. If the action can be integrated to a free para-holomorphic action of a Lie group A such that the quotient map $M \to \bar{M} := M/A$ is a para-holomorphic submersion, then $\bar{M}$ is called a *projective special para-Kähler manifold*. The quotient $\bar{M}$ carries a canonical para-Kähler metric $\bar{g}$ compatible with the induced para-complex structure P on $\bar{M} = M/A$.

Next we explain the extrinsic construction of special para-Kähler manifolds. Recall that a *para-complex vector space* V of dimension n is simply a free module $V \cong C^n$ over the ring $C = \mathbb{R}[e]$, $e^2 = 1$, of *para-complex numbers*. Notice that C^n is a para-complex manifold with the para-complex structure $v \mapsto ev$ and any para-complex manifold of real dimension $2n$ is locally isomorphic to C^n. An $\mathbb{R}$-linear map $\tau : V \to V$ on a para-complex vector space is called *anti-linear* if $\tau(ev) = -e\tau(v)$ for all $v \in V$. An example is the para-complex conjugation $C^n \to C^n, z = x+ey \mapsto \bar{z} := x-ey$. Let (V, Ω) be a para-complex symplectic vector space of dimension $2n$ endowed with a real structure (i.e. an anti-linear involution) τ such that (4.47) holds true. Then

$$\gamma := e\Omega(\cdot, \tau\cdot) \tag{4.50}$$

is a para-Hermitian form and $g_V := \operatorname{Re} \gamma$ is a flat para-Kähler metric on V.

Definition 4.4.11 A para-holomorphic immersion $\phi : M \to V$ from a para-complex manifold (M, P) of real dimension 2n into V is called

(i) *nondegenerate* if $\phi^*\gamma$ is nondegenerate,
(ii) *Lagrangian* if $\phi^*\Omega = 0$ and
(iii) *conical* if $\phi(p) \in d\phi(T_pM)$ and $\gamma(\phi(p), \phi(p)) \neq 0$ for all $p \in M$.

Theorem 4.4.12 ([44])

(i) *Any nondegenerate para-holomorphic Lagrangian immersion $\phi : M \to V$ induces on the para-complex manifold (M, P) the structure of a special para-Kähler manifold (M, P, g, ∇), where $g = \operatorname{Re} \phi^*\gamma$ and ∇ is determined by the condition $\nabla\phi^*\alpha = 0$ for all $\alpha \in V^*$ which are real valued on V^τ.*
(ii) *Any conical nondegenerate para-holomorphic Lagrangian immersion $\phi : M \to V$ induces on (M, P) the structure of a conical special para-Kähler manifold (M, P, g, ∇, ξ). The vector field ξ is determined by the condition $d\phi\,\xi(p) = \phi(p)$.*

The (Affine) Special Para-Kähler Manifold as a Para-Complex Lagrangian Cone

Now we consider the real symplectic module $V_0 = \wedge^3\mathbb{R}^6$ of $G = \mathrm{SL}(6, \mathbb{R})$. For convenience, the standard basis of $\mathbb{R}^6$ is denoted by $(e_1, e_2, e_3, f_1, f_2, f_3)$. The para-complexification $V := V_0 \otimes C = \wedge^3 C^6 \cong C^{20}$ of V_0 is a para-complex symplectic vector space endowed with a real structure τ such that $V^\tau = V_0$ and (4.47). We put $u_i := e_i + ef_i$ and consider the orbit

$$V \subset M = \mathrm{GL}^+(6, \mathbb{R})p \cong \mathrm{GL}^+(6, \mathbb{R})/\mathrm{SL}(3, \mathbb{R}) \times \mathrm{SL}(3, \mathbb{R})$$

of the element $p = u_1 \wedge u_2 \wedge u_3$.

Theorem 4.4.13 $M = \mathrm{GL}^+(6,\mathbb{R})p \subset V$ *is a nondegenerate para-complex Lagrangian cone. The inclusion* $M \subset V$ *induces on* M *an* $\mathrm{SL}(6,\mathbb{R})$*-invariant special para-Kähler structure. The image* $\bar{M} = \pi(M) \cong \mathrm{SL}(6,\mathbb{R})/\mathrm{S}(\mathrm{GL}(3,\mathbb{R})\times\mathrm{GL}(3,\mathbb{R}))$ *under the projection* $\pi : V' \to P(V')$ *is a homogeneous projective special para-Kähler manifold of real dimension 18. Here* $V' \subset V$ *stands for the subset of nonisotropic vectors.*

Proof Using the formulas (4.43) and (4.50) with $\nu = eu_{123} \wedge \bar{u}_{123} = -8e_{123} \wedge f_{123}$ we compute: $\gamma(p,p) = 1$. This shows that $M = \mathrm{GL}^+(6,\mathbb{R})p \subset V'$ consists of spacelike vectors. The tangent $T_pM \subset V$ has the following basis:

$$(u_{123}, \bar{u}_1 \wedge u_{23}, \bar{u}_2 \wedge u_{31}, \bar{u}_3 \wedge u_{12}, \bar{u}_2 \wedge u_{23}, \bar{u}_2 \wedge u_{12}, \bar{u}_3 \wedge u_{23}, \bar{u}_1 \wedge u_{13}, \bar{u}_3 \wedge u_{13}, -\bar{u}_1 \wedge u_{12}).$$

The restriction of Ω to T_pM is zero in view of (4.43). The para-Hermitian form $\gamma|_{T_pM}$ is represented by the matrix (4.49). This shows that $\gamma|_{T_pM}$ is nondegenerate. Hence, the inclusion $M \subset V$ is a conical para-holomorphic nondegenerate Lagrangian immersion. In virtue of Theorem 4.4.12 it induces an $\mathrm{SL}(6,\mathbb{R})$-invariant conical special para-Kähler structure (P, g, ∇, ξ) on M, which in turn induces a homogeneous projective special para-Kähler structure on $\bar{M} = \pi(M) \subset P(V')$. $\square$

The Special Para-Kähler Manifold as an Open Orbit of $\mathrm{GL}^+(6,\mathbb{R})$ on $\wedge^3\mathbb{R}^6$

The conical special Kähler manifold $M = \mathrm{GL}^+(6,\mathbb{R})/\mathrm{SL}(3,\mathbb{C})$ described in Theorem 4.4.7 as a complex Lagrangian cone $M \subset V_0 \otimes \mathbb{C}$ can be identified with the open $\mathrm{GL}^+(6,\mathbb{R})$-orbit $\{\lambda < 0\} \subset V_0 = \wedge^3\mathbb{R}^6$, where λ stands for the quartic $\mathrm{SL}(6,\mathbb{R})$-invariant, see Eq. (2.4):

Proposition 4.4.14 *The projection* $\rho : V_0 \otimes \mathbb{C} \to V_0$, $v \mapsto \mathrm{Re}\,v$, *induces a* $\mathrm{GL}^+(6,\mathbb{R})$*-equivariant diffeomorphism from the Lagrangian cone* $C(X_{(0)}) \subset V_0 \otimes \mathbb{C}$ *described in Theorem 4.4.7 onto* $\{\lambda < 0\} \subset V_0$:

$$C(X_{(0)}) \cong \{\lambda < 0\} \cong \mathrm{GL}^+(6,\mathbb{R})/\mathrm{SL}(3,\mathbb{C}).$$

Proof This follows from the fact that λ is negative on the real part of a non-zero decomposable $(3,0)$-vector, since $\{\lambda < 0\} \cong \mathrm{GL}^+(6,\mathbb{R})/\mathrm{SL}(3,\mathbb{C})$ is connected, see Proposition 2.1.5. $\square$

In that picture the complex structure is less obvious than in the complex Lagrangian picture but the flat connection and symplectic (Kähler) form are simply the given structures of the symplectic vector space V_0. The complex structure is then obtained from the metric, which is the Hessian of the function $f = \sqrt{|\lambda|}$. (We consider λ as a scalar invariant by choosing a generator of $\wedge^6\mathbb{R}^6$.) This route was followed by Hitchin in [76].

The other open $\mathrm{GL}^+(6,\mathbb{R})$-orbit $\{\lambda > 0\} \subset V_0$ cannot be obtained as the real image of a $\mathrm{GL}^+(6,\mathbb{R})$-orbit on the complex Lagrangian cone $C(X) \subset V_0 \otimes \mathbb{C}$ over the highest weight variety $X \subset P(V_0 \otimes \mathbb{C})$. In fact, $GL^+(6, \mathbb{R})$ has only one open orbit on X, see Theorem 4.4.7, and that orbit maps to $\{\lambda < 0\} \subset V_0$ under the projection $V_0 \otimes \mathbb{C} \to V_0$. Instead we have:

Proposition 4.4.15 *The projection* $\rho : V_0 \otimes C \to V_0$, $v \mapsto \mathrm{Re}\, v = \frac{v + \bar{v}}{2}$, *induces a* $GL^+(6, \mathbb{R})$-*equivariant diffeomorphism from the para-complex Lagrangian cone* $M = GL^+(6, \mathbb{R})p \subset V_0 \otimes C$, $p = u_{123}$, *described in Theorem 4.4.13 onto the open* $GL^+(6, \mathbb{R})$-*orbit* $\{\lambda > 0\} \subset V_0$:

$$M \cong \{\lambda > 0\} \cong GL^+(6, \mathbb{R}) / (SL(3, \mathbb{R}) \times SL(3, \mathbb{R})).$$

Proof It suffices to check that $\mathrm{Re}\, u_{123} \in \{\lambda > 0\}$. This follows from the expression

$$\begin{aligned}
2\mathrm{Re}\, u_{123} &= ((e_1 + ef_1) \wedge (e_2 + ef_2) \wedge (e_3 + ef_3) \\
&\quad + (e_1 - ef_1) \wedge (e_2 - ef_2) \wedge (e_3 - ef_3)) \\
&= ((e_1 + f_1) \wedge (e_2 + f_2) \wedge (e_3 + f_3) \\
&\quad + (e_1 - f_1) \wedge (e_2 - f_2) \wedge (e_3 - f_3)),
\end{aligned}$$

since a three-vector belongs to $\{\lambda > 0\}$ if and only if it can be written as the sum of two decomposable three-vectors which have a non-trivial wedge product, see Proposition 2.1.5.   □

Let us denote by ∇ the standard flat connection of the vector space V_0, by ξ the position vector field, by ω its $SL(6, \mathbb{R})$-invariant symplectic form and by X_f the Hamiltonian vector field associated to the function $f = \sqrt{\lambda}$. Then we have:

Theorem 4.4.16 *The data* $(P = \nabla X_f, g = \omega \circ P, \nabla, \xi)$ *define on* $U = \{\lambda > 0\} \subset V_0$ *an* $SL(6, \mathbb{R})$-*invariant conical special para-Kähler structure.*

Proof Any three-vector $\psi \in U$ can be written uniquely as $\psi^+ + \psi^-$ with decomposable three-vectors $\psi^\pm$ such that $\psi^+ \wedge \psi^- = f(\psi)v$, cf. Eq. (2.2) and Corollary 2.1.7. Differentiation at ψ in direction of a vector $\xi \in V_0$ yields

$$(df_\psi \xi)v = (\psi^+ - \psi^-) \wedge \xi = \omega(\psi^+ - \psi^-, \xi)v,$$

that is

$$X_f(\psi) = \psi^+ - \psi^-. \tag{4.51}$$

Using this equation, we can calculate $P = \nabla X_f$ by ordinary differentiation in the vector space V_0. The result is that P acts as identity on the subspace $\wedge^3 E_+ \oplus \wedge^2 E_+ \wedge E_- \subset V_0 = \wedge^3 \mathbb{R}^6$ and as minus identity on the subspace $\wedge^3 E_- \oplus \wedge^2 E_- \wedge E_+$ where $E_\pm = \mathrm{span}\{\alpha \lrcorner \psi^\pm \mid \alpha \in \wedge^2(\mathbb{R}^6)^*\}$ denotes the support of the three-vectors ψ^+ and ψ^-. This shows that $P^2 = \mathrm{Id}$ and that P is skew-symmetric with respect to ω. To prove that the data $(P, g = \omega \circ P, \nabla, \xi)$ define on $U = \{\lambda > 0\} \subset V_0$ an $SL(6, \mathbb{R})$-

invariant conical special para-Kähler structure, it suffices to show that under the map $\rho : V_0 \otimes C \to V_0$ these data correspond to the conical special para-Kähler structure on $M = \mathrm{GL}^+(6, \mathbb{R})p \subset V_0 \otimes C, p = u_{123}$, described in Theorem 4.4.13. It follows from Proposition 4.4.15 and the definition of the structures on M that the data (ω, ∇, ξ) on U correspond to the symplectic structure, flat connection and Euler vector field of the conical special para-Kähler manifold M. One can check by a simple direct calculation that the endomorphism P on $T_{\rho(p)}U$ corresponds to multiplication by $e \in C$ on $T_p M \subset V_0 \otimes C$. This proves the theorem.

Alternatively, we give now a direct argument which avoids the use of Theorem 4.4.13. The structure P on U satisfies

$$d^\nabla P = d^\nabla \nabla X_f = (d^\nabla)^2 X_f = 0,$$

since ∇ is flat. This easily implies the integrability of P by expanding the brackets in the Nijenhuis tensor using that ∇ has zero torsion. In view of the fact that P is skew-symmetric for ω, we conclude that $(U, P, g = \omega \circ P)$ is para-Kähler. Finally, the flat torsion-free connection ∇ satisfies not only $d^\nabla P = 0$ but also $\nabla \omega = 0$, since the two-form ω on V_0 is constant. This proves that (U, P, g, ∇) is special para-Kähler. Now we check that (U, P, g, ∇, ξ) is a conical special para-Kähler manifold, that is $\nabla \xi = D\xi = \mathrm{Id}$. It is clear that $\nabla \xi = \mathrm{Id}$, since ξ is the position vector field in V_0. To prove the second equation, we first remark that the Levi-Civita connection is given by

$$D = \nabla + \frac{1}{2}P\nabla P.$$

(It suffices to check that D is metric and torsion-free.). Therefore, the equation $D\xi = \mathrm{Id}$ is reduced to $\nabla_\xi P = 0$. Let us first prove that ξ is para-holomorphic, that is $L_\xi P = 0$. By homogeneity of f and ω, we have the Lie derivatives

$$L_\xi f = 2f, \quad L_\xi df = 2df, \quad L_\xi \omega = 2\omega, \quad L_\xi \omega^{-1} = -2\omega^{-1}$$

and, hence,

$$L_\xi X_f = 0.$$

The latter equation implies

$$L_\xi P = L_\xi (\nabla X_f) = 0,$$

since ξ is an affine (and even linear) vector field. Using $\nabla_\xi - L_\xi = \nabla \xi = \mathrm{Id}$ we get that

$$\nabla_\xi P = L_\xi P + [\mathrm{Id}, P] = 0.$$

$\square$

References

1. I. Agricola, J. Höll, Cones of G manifolds and Killing spinors with skew torsion. Ann. Mat. Pura Appl. **194**, 673–718 (2015)
2. V. Apostolov, S. Salamon, Kähler reduction of metrics with holonomy G_2. Commun. Math. Phys. **246**(1), 43–61 (2004). doi:10.1007/s00220-003-1014-2. MR2044890 (2005e:53070)
3. D.V. Alekseevsky, Classification of quaternionic spaces with transitive solvable group of motions. Izv. Akad. Nauk SSSR Ser. Mat. **39**(2), 315–362 (1975). 472 [Russian]. MR0402649 (53 #6465)
4. D.V. Alekseevsky, V. Cortés, Classification of stationary compact homogeneous special pseudo-Kähler manifolds of semisimple groups. Proc. Lond. Math. Soc. (3) **81**(1), 211–230 (2000). doi: 10.1112/S0024611500012363. MR1758493 (2001h:53063)
5. D. Alekseevsky, V. Cortés, The twistor spaces of a para-quaternionic Kähler manifold. Osaka J. Math. **45**(1), 215–251 (2008). MR2416658 (2009g:53072)
6. D.V. Alekseevsky, V. Cortés, C. Devchand, Special complex manifolds. J. Geom. Phys. **42**(1–2), 85–105 (2002). doi:10.1016/S0393-0440(01)00078-X. MR1894078 (2003i:53064)
7. D.V. Alekseevsky, N. Blažić, V. Cortés, S. Vukmirović, A class of Osserman spaces. J. Geom. Phys. **53**(3), 345–353 (2005) doi:10.1016/j.geomphys.2004.07.004. MR2108535 (2005h:53067)
8. D.V. Alekseevsky, V. Cortés, A.S. Galaev, T. Leistner, Cones over pseudo-Riemannian manifolds and their holonomy. J. Reine Angew. Math. **635**, 23–69 (2009). doi:10.1515/CRELLE.2009.075. MR2572254 (2011b:53115)
9. D.V. Alekseevsky, K. Medori, A. Tomassini, Homogeneous para-Kählerian Einstein manifolds. Uspekhi Mat. Nauk **64**(1)(385), 3–50 (2009). doi:10.1070/RM2009v064n01ABEH004591 (Russian, with Russian summary) English translation, Russian Math. Surv. **64**(1), 1–43 (2009). MR2503094 (2010k:53068)
10. D.V. Alekseevsky, V. Cortés, T. Mohaupt, Conification of Kähler and hyper-Kähler manifolds. Commun. Math. Phys. **324**2, 637–655 (2013). doi:10.1007/s00220-013-1812-0. MR3117523
11. D.V. Alekseevsky, B.S. Kruglikov, H. Winther, Homogeneous almost complex structures in dimension 6 with semi- simple isotropy. Ann. Global Anal. Geom. **46**(4), 361–387 (2014). doi:10.1007/s10455-014-9428-y. MR3276441
12. C. Bär, Real Killing spinors and holonomy. Commun. Math. Phys. **154**(3), 509–521 (1993). MR1224089 (94i:53042)
13. O. Baues, Prehomogeneous affine representations and flat pseudo-Riemannian manifolds, in *Handbook of Pseudo- Riemannian Geometry and Supersymmetry*. IRMA Lectures in Mathematics and Theoritical Physics, vol. 16 (European Mathematical Society, Zürich, 2010), pp. 731–817. doi:10.4171/079-1/22. MR2681607 (2012a:53082)

© Springer International Publishing AG 2017

L. Schäfer, *Nearly Pseudo-Kähler Manifolds and Related Special Holonomies*,

Lecture Notes in Mathematics 2201, DOI 10.1007/978-3-319-65807-0

14. O. Baues, V. Cortés, Proper affine hyperspheres which fiber over projective special Kähler manifolds. Asian J. Math. **7**(1), 115–132 (2003). MR2015244 (2005a:53076)

15. C.-L. Bejan, Some examples of manifolds with hyperbolic structures. Rend. Mat. Appl. (7) **14**(4), 557–565 (1994) [English, with English and Italian summaries]. MR1312817 (95k:53042)

16. F. Belgun, A. Moroianu, Nearly Kähler 6-manifolds with reduced holonomy. Ann. Global Anal. Geom. **19**(4), 307–319 (2001). doi:10.1023/A:1010799215310. MR1842572 (2002f:53083)

17. L. Bérard Bergery, A. Ikemakhen, Sur l'holonomie des variétés pseudo-riemanniennes de signature $(n \quad n)$. Bull. Soc. Math. France **125**(1), 93–114 (1997) [French, with English and French summaries]. MR1459299 (98m:53087)

18. A.L. Besse, *Einstein Manifolds*. Ergebnisse der Mathematik und ihrer Grenzgebiete (3) [Results in Mathematics and Related Areas (3)], vol. 10 (Springer, Berlin, 1987). MR867684 (88f:53087)

19. C. Böhm, M.M. Kerr, Low-dimensional homogeneous Einstein manifolds. Trans. Am. Math. Soc. **358**(4), 1455–1468 (2006). doi:10.1090/S0002-9947-05-04096-1. MR2186982 (2006g:53056)

20. C. Boyer, K. Galicki, *Sasakian Geometry*. Oxford Mathematical Monographs (Oxford University Press, Oxford, 2008). MR2382957 (2009c:53058)

21. N. Blažić, Paraquaternionic projective space and pseudo-Riemannian geometry. Publications de l'Institut Mathématique Nouvelle série **60**, 101–107 (1996)

22. E. Bonan, Sur des variétés riemanniennes à groupe d'holonomie G_2ou spin (7). C. R. Acad. Sci. Paris Sér. A-B **262**, A127–A129 (1966) [French]. MR0196668 (33 #4855)

23. R.L. Bryant, Metrics with exceptional holonomy. Ann. Math. (2) **126**(3), 525–576 (1987). doi:10.2307/1971360. MR916718 (89b:53084)

24. R.L. Bryant, On the geometry of almost complex 6-manifolds. Asian J. Math. **10**(3), 561–605 (2006). doi:10.4310/AJM.2006.v10.n3.a4. MR2253159 (2007g:53029)

25. J.-B. Butruille, Classification des variétés approximativement kähleriennes homogénes. Ann. Global Anal. Geom. **27**(3), 201–225 (2005). doi: 10.1007/s10455-005-1581-x (French, with English summary). MR2158165 (2006f:53060)

26. J.-B. Butruille, Homogeneous nearly Kähler manifolds, in *Handbook of Pseudo-Riemannian Geometry and Supersymmetry*. IRMA Lectures in Mathematics and Theoretical Physics, vol. 16 (European Mathematical Society, Zürich, 2010), pp. 399–423, doi: 10.4171/079-1/11. MR2681596 (2011m:53083)

27. F.M. Cabrera, SU(3)-structures on hypersurfaces of manifolds with G_2-structure. Monatsh. Math. **148**(1), 29–50 (2006). doi:10.1007/s00605-005-0343-y. MR2229065 (2007b:53059)

28. E. Calabi, Construction and properties of some 6-dimensional almost complex manifolds. Trans. Am. Math. Soc. **87**, 407–438 (1958). MR0130698 (24 #A558)

29. B. Cappelletti-Montano, A. Carriazo, V. Martín-Molina, Sasaki-Einstein and paraSasaki-Einstein metrics from $(\kappa g \quad \mu g$-structures. J. Geom. Phys. **73**, 20–36 (2013). doi:10.1016/j.geomphys.2013.05.001. MR3090100

30. S. Cecotti, C. Vafa, Topological-anti-topological fusion. Nucl. Phys. B **367**(2), 359–461 (1991). doi:10.1016/0550-3213(91)90021-O. MR1139739 (93a:81168)

31. S.G. Chiossi, A. Fino, Conformally parallel G_2structures on a class of solvmanifolds. Math. Z. **252**(4), 825–848 (2006). doi: 10.1007/s00209-005-0885-7. MR2206629 (2007a:53098)

32. S. Chiossi, S. Salamon, The intrinsic torsion of SU(3) and G_2structures, in *Differential Geometry, Valencia, 2001* (World Scientific Publisher, River Edge, NJ, 2002), pp. 115–133. MR1922042 (2003g:53030)

33. S.G. Chiossi, A. Swann, G_2-structures with torsion from half-integrable nilmanifolds. J. Geom. Phys. **54**(3), 262–285 (2005). doi:10.1016/j.geomphys.2004.09.009. MR2139083 (2006a:53054)

34. R. Cleyton, A. Swann, Einstein metrics via intrinsic or parallel torsion. Math. Z. **247**(3), 513–528 (2004). doi:10.1007/s00209-003-0616-x. MR2114426 (2005i:53054)

35. D. Conti, T.B. Madsen, The odd side of torsion geometry. Ann. Mat. Pura Appl. (4) **193**(4), 1041–1067 (2014). doi:10.1007/s10231-012-0314-6. MR3237915

36. D. Conti, S. Salamon, Generalized Killing spinors in dimension 5. Trans. Am. Math. Soc. **359**(11), 5319–5343 (2007). doi: 10.1090/S0002-9947-07-04307-3. MR2327032 (2008h:53077)

37. D. Conti, A. Tomassini, Special symplectic six-manifolds. Q. J. Math. **58**(3), 297–311 (2007). doi:10.1093/qmath/ham013. MR2354920 (2008m:53118)

38. V. Cortés, Alekseevskian spaces. Differ. Geom. Appl. **6**(2), 129–168 (1996). doi: 10.1016/0926- 2245(96)89146-7. MR1395026 (97m:53079)

39. V. Cortés, Odd Riemannian symmetric spaces associated to four-forms. Math. Scand. **98**(2), 201–216 (2006). MR2243702 (2007f:53057)

40. V. Cortés, L. Schäfer, Topological-antitopological fusion equations, pluriharmonic maps and special Kähler mani- folds. Progress in Mathematics, vol. 234 (Birkhäuser Boston, Boston, MA, 2005), pp. 59–74

41. V. Cortés, L. Schäfer, Flat nearly Kähler manifolds. Ann. Global Anal. Geom. **32**(4), 379–389 (2007). doi:10.1007/s10455- 007-9068-6. MR2346224 (2009a:53048)

42. V. Cortés, L. Schäfer, Differential geometric aspects of the tt*-equations, in *From Hodge Theory to Integrability and TQFT tt*-Geometry*. Proceedings of Symposia in Pure Mathematics, vol. 78 (American Mathematical Society, Providence, RI, 2008), pp. 75–86. doi: 10.1090/pspum/078/2483749. MR2483749 (2010f:53103)

43. V. Cortés, L. Schäfer, Geometric structures on Lie groups with flat bi-invariant metric. J. Lie Theory **19**(2), 423–437 (2009). MR2572138 (2010m:53104)

44. V. Cortés, C. Mayer, T. Mohaupt, F. Saueressig, Special geometry of Euclidean supersymmetry. I. Vector multi- plets. J. High Energy Phys. **3**(028), 73pp. (2004) [electronic] doi:10.1088/1126-6708/2004/03/028. MR2061551 (2005c:53055)

45. V. Cortés, M.-A. Lawn, L. Schäfer, Affine hyperspheres associated to special para-Kähler manifolds. Int. J. Geom. Methods Mod. Phys. **3**(5–6), 995–1009 (2006). doi:10.1142/S0219887806001569. MR2264401 (2007j:53077)

46. V. Cortés, T. Leistner, L. Schäfer, F. Schulte-Hengesbach, Half-flat structures and special holonomy. Proc. Lond. Math. Soc. (3) **102**(1), 113–158 (2011). doi:10.1112/plms/pdq012. MR2747725 (2012d:53075)

47. V. Cruceanu, P. Fortuny, P.M. Gadea, A survey on paracomplex geometry. Rocky Mt. J. Math. **26**(1), 83–115 (1996). doi:10.1216/rmjm/1181072105. MR1386154 (97c:53112)

48. B. Dubrovin, Geometry and integrability of topological-antitopological fusion. Commun. Math. Phys. **152**(3), 539–564 (1993). MR1213301 (95a:81227)

49. M.J. Duff, B.E.W. Nilsson, C.N. Pope, Kaluza-Klein supergravity. Phys. Rep. **130**(1–2), 1–142 (1986). doi:10.1016/0370-1573(86)90163-8. MR822171 (87f:83061)

50. J. Eells, S. Salamon, Twistorial construction of harmonic maps of surfaces into four-manifolds. Ann. Scuola Norm. Sup. Pisa Cl. Sci. (4) **12**(4), 589–640 (1985/1986). MR848842 (87i:58042)

51. N. Ejiri, Totally real submanifolds in a 6-sphere. Proc. Am. Math. Soc. **83**(4), 759–763 (1981). doi:10.2307/2044249. MR630028 (83a:53033)

52. M. Falcitelli, A. Farinola, S. Salamon, Almost-Hermitian geometry. Differ Geom. Appl. **4**(3), 259–282 (1994). doi: 10.1016/0926-2245(94)00016-6. MR1299398 (95i:53047)

53. M. Fernández, S. Ivanov, V. Muñoz, L. Ugarte, Nearly hypo structures and compact nearly Kähler 6-manifolds with conical singularities. J. Lond. Math. Soc. (2) **78**(3), 580–604 (2008). doi:10.1112/jlms/jdn044. MR2456893 (2009m:53061)

54. S. Ferrara, S. Sabharwal, Quaternionic manifolds for type II superstring vacua of Calabi-Yau spaces. Nucl. Phys. B **332**(2), 317–332 (1990). doi:10.1016/0550-3213(90)90097-W. MR1046353 (91g:53051)

55. L. Foscolo, M. Haskins, New G2 holonomy cones and exotic nearly Kaehler structures on the 6-sphere and the product of a pair of 3-spheres. arXiv:1501.07838. https://spiral.imperial.ac.uk:8443/handle/10044/1/49672

56. T. Friedrich, Der erste Eigenwert des Dirac-Operators einer kompakten, Riemannschen Man-nigfaltigkeit nichtnega- tiver Skalarkrümmung. Math. Nachr. **97**, 117–146 (1980) [German] MR600828 (82g:58088)
57. T. Friedrich, S. Ivanov, Parallel spinors and connections with skew-symmetric torsion in string theory. Asian J. Math. **6**(2), 303–335 (2002). MR1928632 (2003m:53070)
58. T. Friedrich, I. Kath, A. Moroianu, U. Semmelmann, On nearly parallel G_2-structures. J. Geom. Phys. **23**(3–4), 259–286 (1997). doi:10.1016/S0393-0440(97)80004-6. MR1484591 (98j:53053)
59. P.M. Gadea, J.M. Masque, Classification of almost para-Hermitian manifolds. Rend. Mat. Appl. (7) **11**(2), 377–396 (1991) [English, with Italian summary]. MR1122346 (92i:53028)
60. S. Gallot, Équations différentielles caractéristiques de la sphére. Ann. Sci. École Norm. Sup. (4) **12**(2), 235–267 (1979). (French). MR543217 (80h:58051)
61. R. Gover, R. Panai, T. Willse, Nearly Kähler geometry and (2 3 5)-distributions via projective holonomy (2014). arxiv:1403.1959
62. A. Gray, Minimal varieties and almost Hermitian submanifolds. Mich. Math. J. **12**, 273–287 (1965). MR0184185 (32 #1658)
63. A. Gray, Vector cross products on manifolds. Trans. Am. Math. Soc. **141**, 465–504 (1969). MR0243469 (39 #4790)
64. A. Gray, Almost complex submanifolds of the six sphere. Proc. Am. Math. Soc. **20**, 277–279 (1969). MR0246332 (39 #7636)
65. A. Gray, Nearly Kähler manifolds. J. Differ. Geom. **4**, 283–309 (1970). MR0267502 (42 #2404)
66. A. Gray, Riemannian manifolds with geodesic symmetries of order 3. J. Differ. Geom. **7**, 343–369 (1972). MR0331281 (48 #9615)
67. A. Gray, The structure of nearly Kähler manifolds. Math. Ann. **223**(3), 233–248 (1976). MR0417965 (54 #6010)
68. A. Gray, L.M. Hervella, The sixteen classes of almost Hermitian manifolds and their linear invariants. Ann. Mat. Pura Appl. (4) **123**, 35–58 (1980). doi: 10.1007/BF01796539. MR581924 (81m:53045)
69. A. Gray, J.A. Wolf, Homogeneous spaces defined by Lie group automorphisms. II. J. Differ. Geom. **2**, 115–159 (1968). MR0236329 (38 #4625b)
70. R. Grunewald, Six-dimensional Riemannian manifolds with a real Killing spinor. Ann. Global Anal. Geom. **8**(1), 43–59 (1990). doi: 10.1007/BF00055017. MR1075238 (92a:58146)
71. J. Gutowski, S. Ivanov, G. Papadopoulos, Deformations of generalized calibrations and compact non-Kähler man- ifolds with vanishing first Chern class. Asian J. Math. **7**(1), 39–79 (2003). MR2015241 (2004j:53069)
72. F.R. Harvey, *Spinors and Calibrations*. Perspectives in Mathematics, vol. 9 (Academic, Boston, MA, 1990). MR1045637 (91e:53056)
73. A. Haydys, HyperKähler and quaternionic Kähler manifolds with S^1-symmetries. J. Geom. Phys. **58**(3), 293–306 (2008). doi: 10.1016/j.geomphys.2007.11.004. MR2394039 (2008m:53113)
74. C. Hertling, tt*geometry, Frobenius manifolds, their connections, and the construction for singularities. J. Reine Angew. Math. **555**, 77–161 (2003). doi: 10.1515/crll.2003.015. MR1956595 (2005f:32049)
75. N. Hitchin, The geometry of three-forms in six dimensions. J. Differ. Geom. **55**(3), 547–576 (2000). MR1863733 (2002m:53070)
76. N. Hitchin, Stable forms and special metrics, (Bilbao, 2000). Contemporary Mathe-matics, vol. 288 (American Mathematical Society, Providence, RI, 2001), pp. 70–89. doi:10.1090/conm/288/04818. MR1871001 (2003f:53065)
77. Z.H. Hou, On totally real submanifolds in a nearly Kähler manifold. Port. Math. (N.S.) **58**(2), 219–231 (2001). MR1836264 (2002b:53097)

78. S. Ivanov, S. Zamkovoy, Parahermitian and paraquaternionic manifolds. Differ. Geom. Appl. **23**(2), 205–234 (2005). doi: 10.1016/j.difgeo.2005.06.002. MR2158044 (2006d:53025)

79. G.R. Jensen, Imbeddings of Stiefel manifolds into Grassmannians. Duke Math. J. **42**(3), 397–407 (1975). MR0375164 (51 #11360)

80. S. Kaneyuki, M. Kozai, Paracomplex structures and affine symmetric spaces. Tokyo J. Math. **8**(1), 81–98 (1985). doi:10.3836/tjm/1270151571. MR800077 (87c:53078)

81. I. Kath, G^* -structures on pseudo-Riemannian manifolds. J. Geom. Phys. **27**(3–4), 155–177 (1998). doi:10.1016/S0393-0440(97)00073-9. MR1645016 (99i:53023)

82. I. Kath, Killing Spinors on Pseudo-Riemannian Manifolds. Habilitationsschrift an der Humboldt-Universität zu Berlin (1999)

83. G. Ketsetzis, S. Salamon, Complex structures on the Iwasawa manifold. Adv. Geom. **4**(2), 165–179 (2004). doi: 10.1515/advg.2004.012. MR2055676 (2005g:22008)

84. T. Kimura, *Introduction to Prehomogeneous Vector Spaces*. Translations of Mathematical Monographs, vol. 215 (American Mathematical Society, Providence, RI, 2003). Translated from the 1998 Japanese original by Makoto Nagura and Tsuyoshi Niitani and revised by the author. MR1944442 (2003k:11180)

85. T. Kimura, M. Sato, A classification of irreducible prehomogeneous vector spaces and their relative invariants. Nagoya Math. J. **65**, 1–155 (1977). MR0430336 (55 #3341)

86. V.F. Kirichenko, Generalized Gray-Hervella classes and holomorphically projective transformations of gen- eralized almost Hermitian structures. Izv. Ross. Akad. Nauk Ser. Mat. **69**(5), 107–132 (2005). doi: 10.1070/IM2005v069n05ABEH002283 (Russian, with Russian summary); English translation: Izv. Math. **69**(5), 963–987 (2005). MR2179416 (2006g:53028)

87. S. Kobayashi, K. Nomizu, *Foundations of Differential Geometry. Volume II*. Interscience Tracts in Pure and Applied Mathematics, No. 15, vols. I/II (Interscience Publishers Wiley, New York/London/Sydney, 1963/1969)

88. H.B. Lawson Jr., M.-L. Michelsohn, *Spin Geometry*. Princeton Mathematical Series, vol. 38 (Princeton University Press, Princeton, NJ, 1989). MR1031992 (91g:53001)

89. C. LeBrun, On complete quaternionic-Kähler manifolds. Duke Math. J. **63**(3), 723–743 (1991). doi: 10.1215/S0012-7094-91-06331-3. MR1121153 (92i:53042)

90. A.J. Ledger, M. Obata, Affine and Riemannian s-manifolds. J. Differ. Geom. **2**, 451–459 (1968). MR0244893 (39 #6206)

91. H.V. Le, L. Schwachhöfer, Lagrangian submanifolds in strictly nearly K"ahler 6-manifolds. arXiv:1408.6433v1

92. P. Libermann, Sur le probléme d'équivalence de certaines structures infinitésimales. Ann. Mat. Pura Appl. (4) **36**, 27–120 (1954) [French]. MR0066020 (16,520c)

93. O. Macia, A. Swann, Twist geometry of the c-map. Commun. Math. Phys. **336**(3), 1329–1357 (2015). doi:10.1007/s00220-015-2314-z. MR3324146

94. A.I. Mal'cev, On a class of homogeneous spaces. Izvestiya Akad. Nauk. SSSR. Ser. Mat. **13**, 9–32 (1949) [Russian]. MR0028842 (10,507d)

95. J. Milnor, Curvatures of left invariant metrics on Lie groups. Adv. Math. **21**(3), 293–329 (1976). MR0425012 (54 #12970)

96. A. Moroianu, P.-A. Nagy, U. Semmelmann, Deformations of nearly Kähler structures. Pac. J. Math. **235**(1), 57–72 (2008). doi: 10.2140/pjm.2008.235.57. MR2379770 (2009b:53038)

97. A. Moroianu, U. Semmelmann, Generalized Killing spinors and Lagrangian graphs. Differ. Geom. Appl. **37**, 141–151 (2014). doi:10.1016/j.difgeo.2014.09.005. MR3277524

98. P.-A. Nagy, On nearly-Kähler geometry. Ann. Global Anal. Geom. **22**(2), 167–178 (2002). doi:10.1023/A:1019506730571. MR1923275 (2003g:53073)

99. P.-A. Nagy, Nearly Kähler geometry and Riemannian foliations. Asian J. Math. **6**(3), 481–504 (2002). MR1946344 (2003m:53043)

100. B. O'Neill, *Semi-Riemannian Geometry*. Pure and Applied Mathematics, vol. 103 (Academic [Harcourt Brace Jovanovich Publishers], New York, 1983). With applications to relativity. MR719023 (85f:53002)

101. M.S. Raghunathan, *Discrete Subgroups of Lie Groups* (Springer, New York, 1972). Ergebnisse der Mathematik und ihrer Grenzgebiete, Band 68. MR0507234 (58 #22394a)

102. F. Raymond, A.T. Vasquez, 3-manifolds whose universal coverings are Lie groups. Topol. Appl. **12**(2), 161–179 (1981). doi: 10.1016/0166-8641(81)90018-3. MR612013 (82i:57011)
103. R. Reyes Carrión, Some special geometries defined by Lie groups. PhD-thesis, Oxford (1993)
104. S. Salamon, *Riemannian Geometry and Holonomy Groups*. Pitman Research Notes in Mathematics Series, vol. 201 (Longman Scientific & Technical, Harlow, 1989); copublished in the United States with Wiley, New York, 1989. MR1004008 (90g:53058)
105. S. Salamon, Complex structures on nilpotent Lie algebras. J. Pure Appl. Algebra **157**(2–3), 311–333 (2001). doi:10.1016/S0022-4049(00)00033-5. MR1812058 (2002g:53089)
106. L. Schäfer, tt*-bundles in para-complex geometry, special para-Kähler manifolds and para-pluriharmonic maps. Differ. Geom. Appl. **24**(1), 60–89 (2006). doi:10.1016/j.difgeo.2005.07.001. MR2193748 (2007c:53080)
107. L. Schäfer, tt*-geometry on the tangent bundle of an almost complex manifold. J. Geom. Phys. **57**(3), 999–1014 (2007). doi:10.1016/j.geomphys.2006.08.004. MR2275206 (2007m:53023)
108. L. Schäfer, On the structure of nearly pseudo-Kähler manifolds. Monatsh. Math. **163**(3), 339–371 (2011). doi:10.1007/s00605-009-0184-1. MR2805878 (2012m:53106)
109. L. Schäfer, Conical Ricci-flat nearly para-Kähler manifolds. Ann. Global Anal. Geom. **45**(1), 11–24 (2014). doi:10.1007/s10455-013-9385-x. MR3152085
110. L. Schäfer, F. Schulte-Hengesbach, Nearly pseudo-Kähler and nearly para-Kähler six-manifolds, in *Handbook of Pseudo-Riemannian Geometry and Supersymmetry*. IRMA Lectures in Mathematics and Theoritical Physics, vol. 16 (European Mathematical Society, Zürich, 2010), pp. 425–453. doi: 10.4171/079-1/12. MR2681597 (2011g:53055)
111. L. Schäfer, K. Smoczyk, Decomposition and minimality of Lagrangian submanifolds in nearly Kähler manifolds. Ann. Global Anal. Geom. **37**(3), 221–240 (2010). doi:10.1007/s10455-009-9181-9. MR2595677 (2011g:53131)
112. F. Schulte-Hengesbach, Half-flat structures on products of three-dimensional Lie groups. J. Geom. Phys. **60**(11), 1726–1740 (2010). doi: 10.1016/j.geomphys.2010.06.012. MR2679416 (2011j:53074)
113. R.W. Sharpe, *Differental Geometry*. Graduate Texts in Mathematics, vol. 166 (Springer, New York, 1997). Cartan's generalization of Klein's Erlangen program; With a foreword by S. S. Chern. MR1453120 (98m:53033)
114. S. Stock, Lifting SU(3)-structures to nearly parallel G$_2$-structures, J. Geom. Phys. **59**(1), 1–7 (2009). doi: 10.1016/j.geomphys.2008.08.003. MR2479256 (2010b:53049)
115. A. Strominger, Superstrings with torsion. Nucl. Phys. B **274**(2), 253–284 (1986). doi: 10.1016/0550- 3213(86)90286-5. MR851702 (87m:81177)
116. F. Tricerri, L. Vanhecke, *Homogeneous Structures on Riemannian Manifolds*. London Mathematical Society Lecture Note Series, vol. 83 (Cambridge University Press, Cambridge, 1983). MR712664 (85b:53052)
117. J.A. Wolf, Complex homogeneous contact manifolds and quaternionic symmetric spaces. J. Math. Mech. **14**, 1033–1047 (1965). MR0185554 (32 #3020)

Index

3-symmetric space
$S^3 \times S^3$, 8
SL$(2, \mathbb{R}) \times$ SL$(2, \mathbb{R})$, 9

almost para-complex structure, 28
almost para-Hermitian manifold, 28
almost pseudo-Hermitian manifold, 28

characteristic para-Hermitian connection, 42
characteristic pseudo-Hermitian connection, 42

$\mathcal{G}_1$-manifold, 42
Gray-Hervella classification, 31

half-flat structure, 48

isometries
of nearly Kähler structures, 100

Killing vector fields
of nearly Kähler structures, 100

nearly half-flat structure, 48
nearly Kähler manifold
$S^3 \times S^3$, 8

nearly Kähler manifold
automorphisms, 100
left-invariant, 103
of constant type, 43
nearly para-Kähler manifold, 41
defined by exterior system, 49
Ricci-flat, 45
nearly pseudo-Kähler manifold, 41
SL$(2, \mathbb{R}) \times$ SL$(2, \mathbb{R})$, 9, 102
defined by exterior system, 49
Nijenhuis tensor, 31
totally skew-symmetric, 41, 42, 46–48

SL$(3, \mathbb{R})$-structure
half-flat, 48
nearly half-flat, 48
SU(p, q)-structure
half-flat, 48
nearly half-flat, 48

torsion
parallel, 43
totally skew-symmetric, 42

U(p, q)-structure, 30
quasi-integrable, 45

© Springer International Publishing AG 2017
L. Schäfer, *Nearly Pseudo-Kähler Manifolds and Related Special Holonomies*,
Lecture Notes in Mathematics 2201, DOI 10.1007/978-3-319-65807-0

LECTURE NOTES IN MATHEMATICS   🐎 Springer

Editors in Chief: J.-M. Morel, B. Teissier;

Editorial Policy

1. Lecture Notes aim to report new developments in all areas of mathematics and their applications – quickly, informally and at a high level. Mathematical texts analysing new developments in modelling and numerical simulation are welcome.

   Manuscripts should be reasonably self-contained and rounded off. Thus they may, and often will, present not only results of the author but also related work by other people. They may be based on specialised lecture courses. Furthermore, the manuscripts should provide sufficient motivation, examples and applications. This clearly distinguishes Lecture Notes from journal articles or technical reports which normally are very concise. Articles intended for a journal but too long to be accepted by most journals, usually do not have this "lecture notes" character. For similar reasons it is unusual for doctoral theses to be accepted for the Lecture Notes series, though habilitation theses may be appropriate.

2. Besides monographs, multi-author manuscripts resulting from SUMMER SCHOOLS or similar INTENSIVE COURSES are welcome, provided their objective was held to present an active mathematical topic to an audience at the beginning or intermediate graduate level (a list of participants should be provided).

   The resulting manuscript should not be just a collection of course notes, but should require advance planning and coordination among the main lecturers. The subject matter should dictate the structure of the book. This structure should be motivated and explained in a scientific introduction, and the notation, references, index and formulation of results should be, if possible, unified by the editors. Each contribution should have an abstract and an introduction referring to the other contributions. In other words, more preparatory work must go into a multi-authored volume than simply assembling a disparate collection of papers, communicated at the event.

3. Manuscripts should be submitted either online at www.editorialmanager.com/lnm to Springer's mathematics editorial in Heidelberg, or electronically to one of the series editors. Authors should be aware that incomplete or insufficiently close-to-final manuscripts almost always result in longer refereeing times and nevertheless unclear referees' recommendations, making further refereeing of a final draft necessary. The strict minimum amount of material that will be considered should include a detailed outline describing the planned contents of each chapter, a bibliography and several sample chapters. Parallel submission of a manuscript to another publisher while under consideration for LNM is not acceptable and can lead to rejection.

4. In general, **monographs** will be sent out to at least 2 external referees for evaluation.

   A final decision to publish can be made only on the basis of the complete manuscript, however a refereeing process leading to a preliminary decision can be based on a pre-final or incomplete manuscript.

   Volume Editors of **multi-author works** are expected to arrange for the refereeing, to the usual scientific standards, of the individual contributions. If the resulting reports can be forwarded to the LNM Editorial Board, this is very helpful. If no reports are forwarded or if other questions remain unclear in respect of homogeneity etc, the series editors may wish to consult external referees for an overall evaluation of the volume.

5. Manuscripts should in general be submitted in English. Final manuscripts should contain at least 100 pages of mathematical text and should always include

   – a table of contents;
   – an informative introduction, with adequate motivation and perhaps some historical remarks: it should be accessible to a reader not intimately familiar with the topic treated;
   – a subject index: as a rule this is genuinely helpful for the reader.
   – For evaluation purposes, manuscripts should be submitted as pdf files.

6. Careful preparation of the manuscripts will help keep production time short besides ensuring satisfactory appearance of the finished book in print and online. After acceptance of the manuscript authors will be asked to prepare the final LaTeX source files (see LaTeX templates online: https://www.springer.com/gb/authors-editors/book-authors-editors/manuscriptpreparation/5636) plus the corresponding pdf- or zipped ps-file. The LaTeX source files are essential for producing the full-text online version of the book, see http://link.springer.com/bookseries/304 for the existing online volumes of LNM). The technical production of a Lecture Notes volume takes approximately 12 weeks. Additional instructions, if necessary, are available on request from lnm@springer.com.

7. Authors receive a total of 30 free copies of their volume and free access to their book on SpringerLink, but no royalties. They are entitled to a discount of 33.3 % on the price of Springer books purchased for their personal use, if ordering directly from Springer.

8. Commitment to publish is made by a *Publishing Agreement*; contributing authors of multiauthor books are requested to sign a *Consent to Publish form*. Springer-Verlag registers the copyright for each volume. Authors are free to reuse material contained in their LNM volumes in later publications: a brief written (or e-mail) request for formal permission is sufficient.

Addresses:
Professor Jean-Michel Morel, CMLA, École Normale Supérieure de Cachan, France
E-mail: moreljeanmichel@gmail.com

Professor Bernard Teissier, Equipe Géométrie et Dynamique,
Institut de Mathématiques de Jussieu – Paris Rive Gauche, Paris, France
E-mail: bernard.teissier@imj-prg.fr

Springer: Ute McCrory, Mathematics, Heidelberg, Germany,
E-mail: lnm@springer.com

Printed in the United States
By Bookmasters